W9-DIM-498

WITHDRAWN FROM LIBRARY

THE MECHANICAL CLOCK ON THE TOWER OF MUNICH TOWN HALL

PUPPETS AND AUTOMATA

BY

MAX VON BOEHN

TRANSLATED BY

JOSEPHINE NICOLL

WITH A NOTE ON PUPPETS BY

GEORGE BERNARD SHAW

DOVER PUBLICATIONS, INC.

NEW YORK

This Dover edition, first published in 1972, is an unabridged republication of "Part II: Puppets" (including the relevant sections of the Table of Contents, List of Illustrations, Bibliography and Index) of the one-volume English translation *Dolls and Puppets,* originally published by the David McKay Company, Philadelphia [n.d.]. This material corresponds to the second volume, *Puppenspiele,* of the original two-volume German work *Puppen und Puppenspiele,* published by F. Bruckmann, Munich, in 1929.

The Plates, which were printed in color in the Bruckmann and McKay editions, appear in black and white in the present edition.

International Standard Book Number: 0-486-22848-7
Library of Congress Catalog Card Number: 77-189342

Manufactured in the United States of America
Dover Publications, Inc.
180 Varick Street
New York, N.Y. 10014

ACKNOWLEDGMENT

THE author and the publishers wish to thank those in charge of various public and private collections who have given help in the preparation of this book. They feel that they are specially indebted to the following: the Department of Prints at the Staatliche Kunstbibliothek, the Kunstgewerbe Museum, and the Propyläen-Verlag, Berlin; Herr Georg Zink, the town librarian at Heidelberg; Privatdozent Dr Carl Niessen, of Cologne; the Victoria and Albert Museum and the Bethnal Green Museum, London; the Bayerische National-Museum, the Museum für Völkerkunde, the Theater-Museum (Clara-Ziegler Foundation), and the Armee-Museum, Munich; the Germanische National-Museum, the Bayerische Landesgewerbe Anstalt, Nürnberg; the Staatliche Porzellanmanufaktur, Nymphenburg; M. Henri d'Allemagne, of Paris; the Spielzeug-Museum, Sonneberg; the Kunstgewerbe Museum, Zürich.

Herr Dr Lutz Weltmann, of Berlin, was good enough to allow the use of his literary material for a history of the puppet theatre; for this both the author and the publishers welcome the opportunity of offering him their particular thanks. Dr Weltmann's studies were directed principally toward the literary significance of the puppet theatre, and that subject could not have been introduced into this book without making it inordinately lengthy. It is sincerely to be hoped that Dr Weltmann may have the opportunity of bringing before the public his valuable researches.

Grateful acknowledgment is also due to Herr Direktor Dr Glaser, of Berlin, and Herr Geheimrat Dr Schnorr von Carolsfeld, of Munich, for their courtesy and assistance in providing access to the collections under their care.

NOTE ON PUPPETS

By GEORGE BERNARD SHAW

[In the original German a translation of a letter sent by Mr Shaw to Vittorio Podrecca appears in the text after the quotation from Eleonora Duse (p. 395). Feeling that in its passage through two or three languages back to English the ideas might have suffered, I sent my literal rendering together with the German translation to Mr Shaw, who, declaring that he could not now "recapture the original wording" of the letter, very generously sent me a modified version, with a comment to say that he had originally written to Podrecca giving it as his view that "flesh-and-blood actors can learn a great deal about their art from puppets, and that a good puppet-show should form part of the equipment of every academy of stage art." Since the passage printed here has not the form and wording of the letter once sent to Podrecca, and in view of its great importance, I thought it best to abstract it from the position it occupied in Herr von Boehn's book and print it in this place.—*Translator.*]

I always hold up the wooden actors as instructive object-lessons to our flesh-and-blood players. The wooden ones, though stiff and continually glaring at you with the same overcharged expression, yet move you as only the most experienced living actors can. What really affects us in the theatre is not the muscular activities of the performers, but the feelings they awaken in us by their aspect; for the imagination of the spectator plays a far greater part there than the exertions of the actors. The puppet is the actor in his primitive form. Its symbolic costume, from which all realistic and historically correct impertinences are banished, its unchanging star, petrified (or rather lignified) in a grimace expressive to the highest degree attainable by the carver's art, the mimicry by which it suggests human gesture in unearthly caricature—these give to its performance an intensity to which few actors can pretend, an intensity which imposes on our imagination like those images in immovable hieratic attitudes on the stained glass of Chartres Cathedral, in which the gaping tourists seem like little lifeless dolls moving jerkily in the draughts from the doors, reduced to sawdusty insignificance by the contrast with the gigantic vitality in the windows overhead.

G. B. S.

CONTENTS

Turkish Shadow-play ("Karagöz")

ILLUSTRATIONS

PLATES

ILLUSTRATIONS IN THE TEXT

PUPPETS

ILLUSTRATIONS

PUPPETS

ILLUSTRATIONS

PUPPETS

PUPPETS AND AUTOMATA

I

AUTOMATA AND MOVABLE IMAGES

By careful moulding in wax and by clothing it in real cloth the doll could be made to assume an almost perfect resemblance to a human being, but before there could be any question of absolute illusion two difficulties had to be surmounted—movement and speech had to be artificially introduced. For centuries men struggled with this problem, which, however, has only comparatively recently found a solution. Now we have our automatic figures which can walk, move, and speak—indeed, when necessary, deliver lengthy orations.

The road which had to be traversed before this goal was reached, starting as it did in dimmest antiquity, was a long and weary one. Two methods there were of providing this creature of art with what might be called free will—one by using human force and the other by using mechanical force, it being necessary, of course, for both to work in concealment. The releasing of the impulse through the human hand, being the simplest, must have been the oldest method. Herodotus records that at the festival of Osiris Egyptian women were in the habit of carrying around images of the god, each of which had an exceptionally large phallus which could be moved up and down by means of a string, while at the processions of Jupiter Ammon twenty-four priests brought in a statue of that deity which, by a movement of its head, indicated the direction in which it desired to be borne, and similar phenomena were recorded from the temple of Heliopolis. Indeed, the priests of every religion have known the value of movable images for influencing the imagination of the credulous. At Roman processions there was carried about a figure named Manducus, which, like our nutcrackers, could open and shut its gigantic under-jaw. The Norse sagas also mention animated divine images, presumably with men placed behind them. Tricks of a similar kind were by no means unknown in the medieval Christian Church. Many were the crucifixes which moved their heads and showed blood oozing from the wounds in their sides, as well as Madonnas which shed tears. These hand-controlled images, which at big feasts played a part in the service, first disappeared from the English churches at the Reformation; in Catholic France they maintained their position in the church up to the seventeenth century. The *mitouries* celebrated at the Assumption of the Virgin, in August, at the church of St Jacques in Dieppe were widely renowned. On such occasions living persons and images performed together, and the angels which flew about flapping their wings and blowing their trumpets aroused great enthusiasm. These spectacles were finally suppressed in the year 1647.

Of such displays the only relics are the gigantic figures used in some processions particularly associated with those territories in Western and Southern Europe down to the Alps which were originally inhabited by the Celts. Cæsar and Posidonius relate that among the Celts sacrifices were made every five years by the Druids, that on these occasions giant images of wickerwork, wood, and straw were set up, filled with men and sacrificial animals, and then burned. At a much later period in the Ile de France similar giant images, dressed as soldiers, were burned after being carried round at religious processions. In Hainaut between 1456 and 1460 giant figures took part in processions organized during plague years; these made their appearance also at Amiens, Metz, Nevers, Orleans, Poitiers, Laon, and Langres. There they were styled *papoires*, and often were accompanied by dragons, likewise set in motion by concealed men. The majority of the towns in Flanders and Brabant possessed such giants, which the people used to dress up as Goliath or Christophorus. At Douai there was the giant Gayant with his wife and children—figures made of wickerwork, with brightly painted wooden heads, the father 21 feet high, the mother 18 to 20, and the children 12 to 15. Ten or twelve persons were required to move the largest. When Matthieu de Montreuil went with Cardinal Mazarin in 1660 to the royal wedding on an island at the mouth of the Bidassoa he saw at San Sebastian, in Spain, seven gigantic figures carried round during the Corpus Christi procession; these were made of wickerwork and painted canvas, and were intended to represent the Moorish kings and their wives. They were so tall that they reached the roofs of the houses, and each was carried by two or three men concealed inside it. Quite recently processions of this kind were held at Pamplona, when gigantic figures, representing Moors or Normans, visited the town hall in festive procession, bowed before the image of the Blessed Virgin, and danced in front of the cathedral. Similar shows are to be found in Sicily, while at Salzburg the giant Samson was still accompanying processions in the second half of last century.

Primitive races have remarkably similar customs. For their masked dances, apparently designed in honour of the dead, the Baining of New Britain, Oceania, prepare gigantic figures on a

FIG. 1. THE DANCE OF GIANTS
Masquerade of Emperor Maximilian I. From the *Freydal*

light cane framework, covered with oxhide and provided with painted faces. These figures may be as much as 45 m. high, and are carried on bamboo poles. Dancers, panting and stamping their feet, support the figure on their dance spears, and when the leader of the throng sinks down after a few steps under the weight of the forepart of the image, which rests on his head, they let it fall to the ground. The onlookers tear it into fragments, taking the pieces home with them as amulets possessed of magical properties.

The delight taken in the giant-show must have been universal. The romantically minded "last of the knights," Emperor Maximilian I, a somewhat pleasure-seeking and frivolous monarch, who took great pains in the planning of his masquerades, once disguised himself and his fellow-dancers as giants, in comparison with whom their partners cut a pretty comic figure, since they looked just like children. Even as late as the last Tsar's coronation, in May 1896, a procession of gigantic images of this kind was organized in the Khodinskoie Plain—a procession which gained a tragic fame because of the terrible catastrophe which occurred on that occasion.

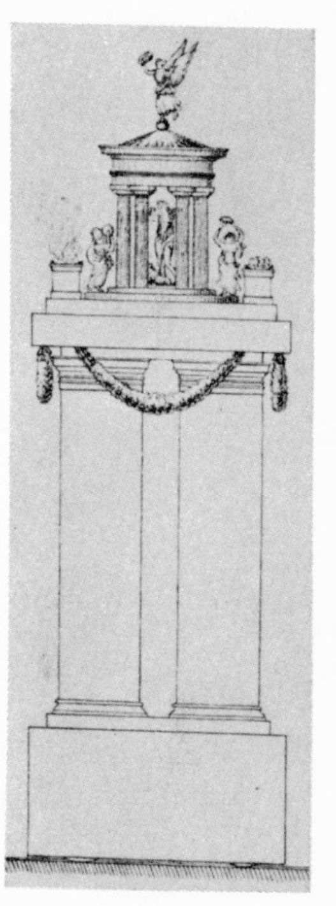

FIG. 2. HERO'S AUTOMATON

After human power as a controlling force comes the use of material substances. These, such as sand, water, and quicksilver, can alter an object's centre of gravity and bring it back once more to its original position. We know that the Chinese and the Greeks were acquainted with this property of quicksilver and utilized it. It is said that by means of quicksilver Dædalus made a wooden image of Venus move. Among the ancients, however, there is no doubt that water played the chief *rôle* as a controlling force; here water-pressure, warmed air, and steam have all to be taken into consideration. The Egyptians in especial were renowned for the construction of automata. At a festival of Bacchus which was held under Ptolemy Philadelphus (285–247 B.C.) there was carried in a figure of the god which dispensed wine from a golden goblet. At that period in the town of Nysa there was an allegorical statue which rose automatically, poured milk from a gold shell, and sat down again. Hero of Alexandria, the most famous mechanician among the ancients, who lived presumably in the second century B.C., has left a treatise concerning automatic theatres in which he describes his own inventions. Wilhelm Schmidt, basing his account on Hero's works, thus describes the movements of a group of his automatic figures. A platform, fitted with three wheels, bearing the apotheosis of Bacchus moved by itself upon a firm, horizontal, and smooth surface up to a certain point and then stopped, at which moment the sacrificial flame burst forth from the altar in front of which Bacchus stood. Milk flowed from his *thyrsus*, and from the goblet he held streamed wine which sprinkled a panther crouched at his feet. Suddenly festoons appeared all round the base of the platform, and figures representing the Bacchantes, to the beating of drums and the clanging of cymbals, danced round the temple within which Bacchus was placed. Then the god turned round to another altar, while a figure of Nike, set on the top of the

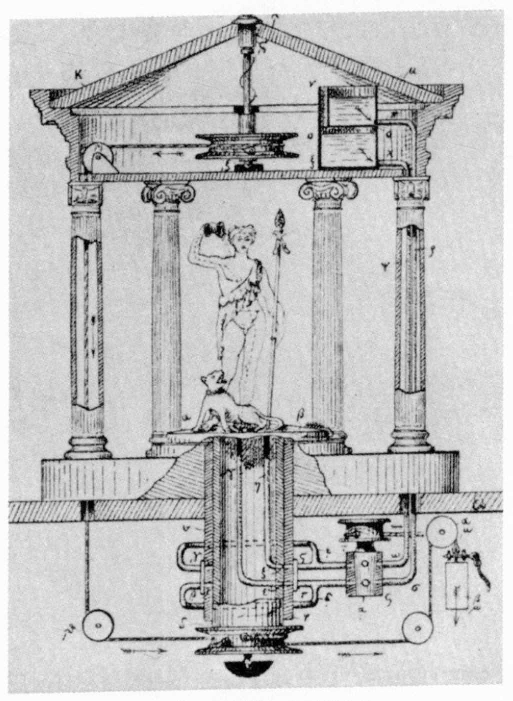

FIG. 3. HERO'S AUTOMATON
After the sketch by Wilhelm Schmidt
Neue Jahrbücher für das klassische Altertum, II, 1899

temple, turned in the same direction. The second altar flamed forth, the *thyrsus* and the goblet flowed again, and the Bacchantes danced. After all these movements had been carried out automatically the platform returned of its own accord to its starting-point.

These automatic shows were carried still further. At the end of the second century B.C. a five-act play, *The Tale of Nauplius*, was produced by Philo of Byzantium in a regular automatic theatre, which had a vertical partition dividing it into machine-room and stage. In the first scene were shown twelve Greeks standing in three rows, as if they were engaged in repairing ships—sawing, cutting timber, boring holes, and hammering. In the second scene appeared the launching of the ships. The third scene displayed a moving landscape. The fleet passed by, with dolphins diving up and down. A storm came on. In the fourth scene Athena made her appearance to Nauplius, who held a torch aloft. The fifth scene introduced the shipwreck; Ajax tried to save himself by swimming, but Athena cast a thunderbolt at him, and he vanished in the waves.

These automata, driven by water-power, were still being made use of during the seventeenth century in the laying-out of gardens. A good example is that of Salomon de Caus in the Schloss park, at Heidelberg, especially famous before the outbreak of the Thirty Years War. Archbishop Markus Sittichus, Count Hohenhembs, got similar automatic playthings introduced into the Hellbrunn park, near Salzburg. These are still to be seen, and continue to give delight to naïve souls. The description which Prince Max of Hohenlohe has given of them is so beautifully expressed that it may be quoted here:

> Some centuries ago the princely magician Markus Sittichus laid out a rare and marvellous garden near Salzburg. In this luxuriant place, shaded by magnificent trees, subterranean streams murmur soft melodies between long alleys and spacious flower-beds and make strange play in magic grottoes. As we step into these moist, dark caverns, all at once a loud warbling of birds, like the song of a thousand nightingales, breaks upon our ears; in the depths small, strange figures begin to move over concealed wells—figures of men and animals, of airy spirits and deities, of clumsy goblins and teasing water-sprites. Water drips from the stone stems of creeping stalactites, and the gleam of all the precious jewels in Aladdin's garden sparkles in the light of the magic lamp from *The Arabian Nights*. And above, over the gurgling grottoes and the singing springs, Markus Sittichus, among other things, built a magnificent pavilion, almost enclosed by an iron curtain.
>
> With what excitement did we children wait for the moving of that curtain in anticipation of the wonders it concealed from us! At last the magic veil was lifted; the curtain rose, and before our enchanted eyes stood, in a small compass, a completely strange world, a theatre of life.
>
> There stood palaces and houses, strange buildings with high, glistening domes and Eastern cupolas, narrow streets, and wide boulevards, all of them filled with little men and women, motionless certainly yet deceitfully lifelike, in their strange, old, multicoloured dresses, and near them horses, cats, and dogs. Before the town hall stood soldiers all ready to march, commanding officers, citizens thronging in rich, heavy garments; in the market-place

merchants were chaffering and drivers speeding. The most extraordinary thing, however, was that we could see, too, through the walls of the houses into the interiors and watch what was going on within—how the bakers baked, how the washer-women washed, how the millers prepared their bright, golden-yellow flour and the cooks got ready their tasty dishes. One could see here the maids sweeping the floors, here a husband and wife fighting with one another in their room, here the children playing, here the teacher giving his lessons from open books, and many things besides.

Suddenly, at a magic knock, a strain of subterranean, mysterious music vibrated from the depths, and at that moment the whole town began to take life, all the figures commenced to move. The officers gave their orders, the soldiers marched off, the citizens really began to crowd along and the merchants to carry on their business. Men ran and speeded by; in the house the bakers started baking, the washer-women washed, and the maids did their ironing. The married couple fought in earnest, the children romped, and the teachers gave their lessons. Everybody, every single one, started moving, really and truly, and lived, and was marvellously real, and the dream was no longer a dream, but genuine pulsating life. The mill creaked and the hens cackled; the girls moved in circles; entertainers and jugglers led fettered bears into the square and made them dance.

Then we children noted with fear that suddenly the clock on the town hall forgot to strike, and all the chimes died away, and one figure after another became automatically slower in movement until it gradually grew rigid. Then all at once everything stood still, and the wonderful music ceased to play. Before we were aware of it the iron curtain sank before our gaze, and with the vanishing of the last houses in the town the strange dream vanished, and all that was true, all that we had believed in, passed again into the realm of the fairy-tale.

The art of making automata was in existence as early as the third century B.C.; small automata must have been known even in Aristotle's time, as his *Physics* proves. Of such objects there is no lack of record. Homer speaks of Vulcan's twenty tripods which automatically entered the banquetting hall of the gods and rolled out again. Archytas of Tarentum, who lived about 390 B.C., is said to have invented a mechanical flying pigeon, but this, of course, is very doubtful. In the great Indian fairy-tale collection of the *Somadeva*, which belongs to the eleventh century, but includes matter of a much earlier date, mention is made of puppets moved by mechanism; one fetched a wreath, another water, a third danced, and a fourth is even said to have spoken. Petronius refers to a silver doll which could move like a living being. Oriental and Byzantine sources are especially rich in their records of such automata. Several caliphs are said to have been in possession of trees on the branches of which mechanical birds sat singing and flapping their wings. The Emperor of the East, Constantine VII Porphyrogenitus, owned not only such a tree, but also a huge, mechanical, roaring lion. Descriptions of these wonders aroused the imaginations of Western poets. Thus the Tristan saga has its automaton, a

FIG. 4. AUTOMATIC TOY OF THE EIGHTEENTH CENTURY
Collection of Henri d'Allemagne, Paris

temple with the figure of Isolde, on whose sceptre sits a bird flapping its wings and at whose feet lies a dog shaking its head. In the Middle Ages these things were very popular. The sketch book of Willars de Honecourt, which belongs to about the year 1245, contains a drawing of an eagle which could move its head when the deacon read the epistle in the church. Albertus Magnus, who died in 1280, is said to have made an automaton—a lovely woman who could speak—which the zealous Thomas Aquinas destroyed, declaring that it was the work of an evil spirit.

Clocks with automatic figures appeared at quite an early date, and these too the poets (for example, in *Der jüngere Titurel*) loved to describe. The huge clock in Strasbourg Cathedral in 1352 showed the three kings bowing and a crowing cock. The so-called *Männleinlaufen* on the clock of the Marienkapelle in Nürnberg, which was constructed between 1356 and 1361, was famous. There the Emperor Charles IV was shown sitting on his throne, and when the clock struck twelve the seven prince electors passed before him and bowed. The man who struck the hours with a hammer on a bell was well known by about 1380, and the clocks at Lubeck, Danzig, Heilbronn, Bern, Ulm, and elsewhere displayed a variety of other mechanical movements—crowing cocks, bears that shook their heads, a king who turned a sand-glass, and the twelve apostles moving in a circle. That these playthings have lost none of their attraction can be denied by no one who at 11 A.M. has watched the mechanism of the great clock which the merchant Karl Rosipal erected on the tower of the new town hall in the Marienplatz at Munich.

Regiomontanus in the fifteenth century is said to have made a fly which fluttered round the room, and similar wonderful inventions are recorded in Spain. Don Enrique de Villena, a reputed magician, made a head which could speak. In the cathedral of Toledo Don Alvaro de Luna, who ended his days on the scaffold in 1453, was reputed to have got tombstones made for himself and his wife which were so artfully contrived that, when Mass was said, the recumbent figures raised themselves and knelt. Isabella the Catholic is supposed to have ordered the destruction of these remarkable mechanisms on the ground that they were unsuitable for the dignity of the place and the solemnity of the occasion. Leonardo da Vinci, who took a keen interest in all sorts of mechanics, constructed in 1509 a lion which could walk about and which could open its breast, to reveal within the lilies of France. Emperor Charles V sought to while away time in his old age at the cloister of Yuste by playing with automata. The Milanese engineer Giovanni Torriani helped him in this; it was he who constructed an automatic wooden male figure which went daily to the archbishop's palace at Toledo and thence fetched bread.

In the wine-loving sixteenth century men liked to use automata as so-called drinking clocks, interesting examples of which are to be found in the Grünes Gewölbe, at Dresden. One of these, a mechanical toy, represents St George engaged in freeing Princess Aja of Lybia from the dragon. It is wound up and allowed to run on its wheels along the table, and it is a rule that one must empty one's glass before it comes to a stop. The other, made by a native of Nürnberg, C. Werner, is a gilded silver centaur engaged in carrying off a woman. When the clock is going the eyes of the two figures move, but if the base of the concealed mechanism is pulled twice the two foremost hounds spring up, the clock runs round in a circle on its wheels, and the centaur shoots at the guests arrows which are taken from his quiver and placed in the bow, and whoever is struck must give proof of his drinking capabilities.

Already in this century automatic devices were included in the equipment which wandering showmen took round to exhibit for money. In 1582 Peter Döpfer, of Schneeberg, displayed in Nürnberg a mechanical mine. In 1587 the town council of Nürnberg gave permission to Daniel Bertel, of Lubeck, "to show for three days at the price of a pfennig his mechanical pieces and toys, as first a fleet of galleys with Turks and Christians fighting on the sea, also merry Swiss and German dances and pretty acrobatics." In November 1611 "two foreign wax-modellers, who have brought here some figures which move by themselves," were allowed to exhibit their wonders for three days in Nürnberg.

Gottfried Hautsch, who died in 1703, was already well known as a maker of tin soldiers when he constructed in Nürnberg a mechanical automaton with many figures, which was nicknamed his "little world." This is a kind of automaton which, to distinguish it from others, is technically indicated by the term *theatrum mundi.* The *theatrum mundi* for centuries provided the traditional afterpiece of the wandering marionette theatres; by means of small movable figures running on rails it showed a diversity of scenes, such as the creation of the world and Noah and the Flood. Flockton's show, which was exhibited in England at the end of the eighteenth century, utilized five hundred figures, all employed in different ways and manners. Here there were movable figures of a peculiarly ingenious kind: swans, for example, which dipped their heads in the water, spread their wings, and craned their long necks to clean their feathers. Even in 1803 Reichardt was expressing admiration in Paris at the mechanical theatre of Pierre, who previously had been settled in Vienna. It was only with the success of the film that these mechanical shows declined, but not so long ago one could still see, as a lively and boisterous afterpiece to a popular puppet-show, the battle of Sedan, the siege of the Taku forts, sledging, storms at sea, etc. In the seventeenth century there was another motion-mechanism likewise connected with the toy doll: this was of the type of the automatic figures in Hainhofer's art cabinet already referred to. In the first years of the eighteenth century Abraham a Sancta Clara, the well-known Viennese preacher, speaks of "dolls so ingeniously contrived that on being pulled, pressed, or wound up they become animated and move by themselves as desired." Corvinus, who wrote under the name of Amaranthes, could describe in 1716 the "costly and ingenious dolls which display *actiones* by means of concealed clockwork"; these were then a speciality of Augsburg and Nürnberg, "which are rapidly filling the entire world with them." Such moving dolls, *la jolie catin*, were shown in France on the streets by itinerant girls.

The eighteenth century, when automata seem to have been

FIG. 5. LA CHARMANTE CATIN
Etching by Bouchardon. From the *Cris de Paris*. About 1750

particularly popular, produced masters of this kind of art. The first was Jacques de Vaucanson, who constructed his automata, which rapidly became world-famous, between 1738 and 1741. One was a flute-player who had a repertoire of twelve selections; still more admired was a life-size duck which waddled and quacked, moved its wings and neck, ate corn, and drank water, then, having digested its food, dropped a mass looking like excrement. It was remarkable, too, for the skilfulness with which its plain copper plumage was put together so as to resemble the lovely shimmering shades of a real duck. These automata were exhibited all over Europe and eventually came into the possession of the well-known aulic councillor Beireis, of Helmstedt, in whose rather disorderly collection Goethe saw them in 1805 in a somewhat dilapidated condition. Feldhaus knew of a similar automaton—a peacock constructed fifty years before by a French general, which could also walk, eat, and digest with palpable success.

The position occupied by Vaucanson was later taken by the two Droz, Pierre Jaquet Droz, the father, and Henri Louis Jaquet Droz, the son, of La Chaux de Fonds. The former in 1760 made a child doll capable of writing, and the latter in 1773 constructed two figures, one of which could draw and the other play the piano. All three are still preserved in the museum at Neuchatel. The third of these skilled inventors is Wolfgang von Kempelen, who in 1769 made a chess-player, which he exhibited with the greatest success; no one then could discover the secret of its mechanism. In 1809 Napoleon played a game of chess with it and lost. Here, however, we are not dealing with a pure automaton, for in the box on which the figure was mounted was concealed a man who, by a system of mirrors, could see the chess-board and the movements made by his opponent. Another speaking figure, this time a real automaton, was made in 1778 by Kempelen. It is curious to observe that Tippoo Sahib, Sultan of Mysore, who had every reason to hate the English, got an automatic group constructed for his pleasure; this included a life-size tiger which rushed roaring to devour an Englishman in uniform. After the death of the Sultan, in 1799, this proof of his sentiments was found among the treasures at Seringapatam.

That to-day we have performing dolls which can not only move, but speak and sing (by means of an inserted gramophone) has already been mentioned in the section dealing with the development of the toy doll. It is not very long ago that a record performance by one of these automata was vouchsafed us. In September 1928 an exhibition of model engines was to have been opened in London by a prominent man. At the last moment he had to decline. A mechanical man, clad like a robber knight, in full armour, took the chair on the platform in his stead. Calmly he waited till the audience assembled; then he rose, took a few steps backward to the speaker's desk, and began to address them. He spoke clearly and precisely, in a tone which carried farther than any human voice could have done. He referred to the wonders of the exhibition and to the future of engineering, putting himself forward as a fitting example of this science. His eyes sparkled, his lips moved a little to reveal sharp teeth, resembling those of a beast of prey. It is reported that a partner is being made for him, a mechanical figure of a woman. The impertinent fancies of *Simplizissimus* concerning mechanical tsars are thus far outdistanced by the reality.

II

THE ORIGIN OF THE PUPPET-SHOW

The child that occupies itself with its doll is playing, without knowing or desiring it, at a theatre, and thereby is poetically and dramatically active. It puts itself in the place of the doll; it is artist and public in one person; it creates the object of art and appreciates it at one and the same time. Herein lies the basis of the puppet theatre, for the first child which played with its doll as if it were a living being may be regarded as the originator of the marionette stage. When and where this happened we naturally do not know, even as little as we know where and at what period adults first perceived this tendency of the child's mind and created an art form out of its naïve play.

Some scholars, the chief representative of whom is Richard Pischel, seek the home of the puppet-show in India. From there it is said to have gone to Persia, and to have reached Europe through the medium of the Arabs. Against this theory it may be justly argued that proofs of the existence of the puppet-show in ancient Greece carry us back to very much earlier times, and that consequently India cannot possibly assume the credit of its invention. When we note the existence of the puppet-show during early centuries spread over all civilized lands, and when we reflect that in truth no great intelligence was required for its invention—indeed, all men had to do was open their eyes and observe their own children—we may think of autochthonic development, and assume that the puppets must all have resulted in general and independently from the preliminary condition—the child's play—which was everywhere existent. Philipp Leibrecht, however, rejects this highly illuminating conception, since he observes that not only the similarity of the persons represented, but also the likeness in the matter pertaining to the shows, testifies undoubtedly to an international connexion. This connexion must be granted, even if we are uncertain as to what way the transition from people to people came about.

It is possible, at least not impossible, that in the puppet-show we have before us the most ancient form of dramatic representation. Certain it is that this puppet-show best harmonized with the intelligence of the people at large, for it came to meet the popular conceptions and appealed to their instincts. On that account the puppet-show is "often a much clearer mirror of the thought and feeling of the people than poetry, and not infrequently is the bearer of ancient traditions." When we try to follow the history of the puppet-show, however, there arises one great difficulty—namely, the lack of precision in the documentary evidence relating to its technique. To-day, following the example of Dr Alfred Lehmann, we classify the puppet-play according to the following plan:

(1) puppet-shows with flat figures: (*a*) the shadow theatre, (*b*) the *theatrum mundi*, (*c*) the theatre of figures.

(2) puppet-shows with rounded figures: (*a*) the hand-puppet theatre, (*b*) the marionette theatre: one worked from below and the other from above.

The records preserved in old texts only rarely provide justification for this division, for the simple reason that the individual author had in his mind only one or the other type of puppet, and for the time being that signified for him *the* puppet-show in general. To this system of classification may be added still another species. Lehmann's grouping takes into consideration only the movable puppets, but we might include also those that are immovable—the kind of show which, when it is represented by human beings, is called *tableaux vivants*. With inanimate figures, this special type may be defined as the mass arrangement of dolls, to which naturally the word *vivants* cannot be applied. The group includes table decorations, Christmas cribs, and tin soldiers, all utilizing a large number of figures.

III

TABLE DECORATIONS AND THE CHRISTMAS CRIB

EVEN in the fifteenth century, in order to add to the pleasures of the table, decoratively constructed dishes of food used to be served up; these, intended for the eye alone, were designed to satisfy the æsthetic and literary pretensions of a society proud of its learning. For making the figures themselves *Tragant*, a composition of flour, water, and paste, was regularly used, the whole group being under the supervision of a skilled *chef* trained in the moulding of forms.

Allegories and symbolic shapes in table decorations made special appeal to the taste of the age. Porcelain was utilized for this purpose in the eighteenth century, and then may be said to have been the first true flourishing time of this type of art. During this period it was a common practice to have the little figures modelled in sets, these sets being intended for arrangement on the table. One of Kändler's masterpieces was just such a table set, called the *Temple of Honour*, which consisted of 123 separate pieces and, when put together, occupied a space 1·16 m. high, 86 cm. wide, and 60 cm. deep. Yet, large as it was, it served merely as the centre piece of a still larger decorative set, in which appeared seventy-four other white figures.

All the porcelain-manufacturers competed in providing magnificent examples of this kind. In 1767 the convention of the Zwettl monastery presented Abbot Reine Kollmann with a set of Vienna ware, 4 m. long and 50 cm. deep, representing the arts, virtues, and crafts. Charles Eugene, Duke of Württemberg, ordered in Ludwigsburg a gigantic table set which showed Neptune with a great retinue of Tritons, naiads, dolphins, fish, etc., the whole built up in a basin of water to be placed in the centre of the table. A table set was part of the service which in 1772 Frederick the Great presented to Catherine II; this displayed the Queen sitting on a canopied throne, completely surrounded by representatives of all the Russian races under her dominion. In 1785 the Prince Palatine, Charles Theodore, presented Cardinal Antonelli with a service of Frankenthal porcelain, which included as table decorations thirteen groups, sixty single figures, forty animals, and thirty-six vases on pedestals. In 1791 a table set of Berlin ware was produced which symbolized nature with all its powers and mysteries in figures of divinities, temples, altars, and obelisks. The confectionary of Count Brühl, the superintendent of which had control of the table decorations, contained not only dozens of figures, but also a number of accessories, such as four churches, two temples, fifty-one town houses, thirteen peasant houses, sheds, stables, niches, gondolas, goats, pyramids, grottoes, reservoirs, cornices, vases, altars, pillars, pedestals, flower-jugs—all designed for use in the building up of table decorations. Even in 1808 Napoleon I, when his friendship with Alexander I was at its closest, ordered for that monarch a Sèvres service in Egyptian style, consisting of temples, obelisks, sphinxes, etc. This set had an advantage over those which had hitherto been usual in that it was capable of being varied, and could be set up, as fancy dictated, now in one way, now in another. The famous table sets of Johann Melchior Dinglinger, in spite of their richness in figures, had not been adaptable in this manner. His most renowned work, representing the Court of the Great Mogul in Delhi and made as a table set for Augustus the Strong, has been called a doll's house, so crowded is it with little gilded and enamelled figures. On this the master himself, together with his family and fourteen apprentices, is said to have worked from 1701 to 1708, and for it he is said to have received 58,485 thalers.

Important in a much deeper sense is the Christmas crib, commonly made up by the utilization of a number of small figures. The crib itself testifies to a genuine, strong religious sentiment, powerful enough to arrest our attention now by its form of expression, while the extant examples of ancient manufacture form valuable and interesting documents for a study of social life in past times. The artists who made these crib-figures took pains to render them as realistic as possible; they accentuated expression, bearing, and gesture so acutely that often they approached caricature. These cribs provide a very effective, even though silent, theatre, so that Rudolf Berliner has every justification for placing the crib-figure alongside the marionette.

The custom of exhibiting a crib at Christmas is a very old one, and those who attribute its invention to St Francis of Assisi are certainly mistaken. That saint, having ever striven to bring closer to the hearts of his contemporaries the human element in the Christian faith, arranged Christmas festivals which approached as nearly as possible to the events in Bethlehem, with a real child, a real manger, and real animals, but the puppet-crib existed long before his time. It is, indeed, as old as the Christmas festival itself, which was first established by Pope Liberius in the year 354. There are extant sermons delivered about the year 400 by St John Chrysostom and St Gregory Thaumaturgus in which references are made to the existence at that period of a crib with figures of the Holy Family, even with figures of an ox and an ass.

FIG. 6. ITALIAN CRIB-FIGURE OF THE EIGHTEENTH CENTURY
From *Denkmäler der Krippenkunst*, by R. Berliner (B. Filser, Augsburg)
Bayerisches National-Museum, Munich

The ancient Church loved such shows, and these produced a much more powerful effect on uneducated folk (probably also on educated people as well) than the Mass and the sermon. The procession with the palm ass has been mentioned elsewhere in another connexion; similar ceremonies were carried out on Good Friday, at Easter and Whitsuntide. The Christmas crib, half a toy and half an object of devotion, just as it is to-day, seems to have its home in Italy and in particular in Naples, where it was to be found in its richest and most artistic forms. The oldest specific record of the crib which Georg Hager can adduce comes from Naples, where in the year 1478 a certain Jaconello Pepe gave a commission to two sculptors, Pietro and Giovanni Alamanno, to make a crib for his family chapel in S. Giovanni a Carbonara. The individual pieces are fully detailed: the Blessed Virgin with a crown, St Joseph, the *Bambino*, eleven angels, two prophets, two sibyls, three shepherds, twelve sheep, two dogs, four trees, an ox, and an ass. In 1507 a crib of twenty-eight figures was ordered from

Pietro Belverte, of Bergamo, for the church of S. Domenico Maggiore in Naples, while in 1558 the nuns of Sapienza commissioned Annibale Caccarello to make a crib of fourteen figures, for which they paid 140 ducats.

By the sixteenth century the crib is to be found on this side of the Alps; an inventory of 1537 relating to the Carmelite monastery in Bruges mentions two cribs, while by the seventeenth century the Jesuits had introduced them to Munich; in 1607 for the first time the fathers installed such a crib in the Michaelshofkirche, and provided for it incidental music. In old Bavaria the crib was very greatly beloved and became extremely popular. The Benedictine nuns at Frauenchiemsee in 1627 obtained their first crib, which was built up as a 'mountain,' so as to provide the greatest possible space for figures, animals, and so on. The crib, like porcelain ware, flourished most in the eighteenth century, receiving special attention in Naples and in Sicily, where a playful temperament, artistic talents, bigotry, and an inexhaustible imagination combined to create the most extraordinary and the most charming specimens of the type. The figures were made by distinguished sculptors, such as Giuseppe Sammartino, who had won such fame through his *Christus* in the chapel of the Princess Sangro, Domenico Antonio Vaccaro, Matteo Bottiglieri, F. Celebrano, Niccolo Somma, Lorenzo Mosca, all of whom were at the same time the modellers of the Capodimonte porcelain ware. The heads were made of terra-cotta, with glass eyes, the hands and feet of wood set on wire limbs, which permitted them to be placed in any desired position. The painting of the skin was carried out on a chalk ground, with an under wash in tempera, and finished off with oil-colour varnish—a technique which not only gave opportunity for a rich variety of flesh colours, but also produced a fine varnished surface. They were constructed according to the rules of perspective, the larger figures of the foreground not exceeding 37 to 40 cm. in height. The bodies of the animals were made of terra-cotta, the legs of lead, the ears of tin, while the buildings for the most part were constructed of wood or cork.

For the dresses of their crib-figures the Neapolitans used real cloth, the seams of which were edged with wire thread so as to secure a more plastic effect; the gold and silver ornaments too were often real. The Sicilian figures were carved from limewood; the dresses were of linen saturated with glue and paste. The material was starched after modelling and then realistically painted. The artists were given the greater scope for their efforts in that a very extensive series of subsidiary themes was

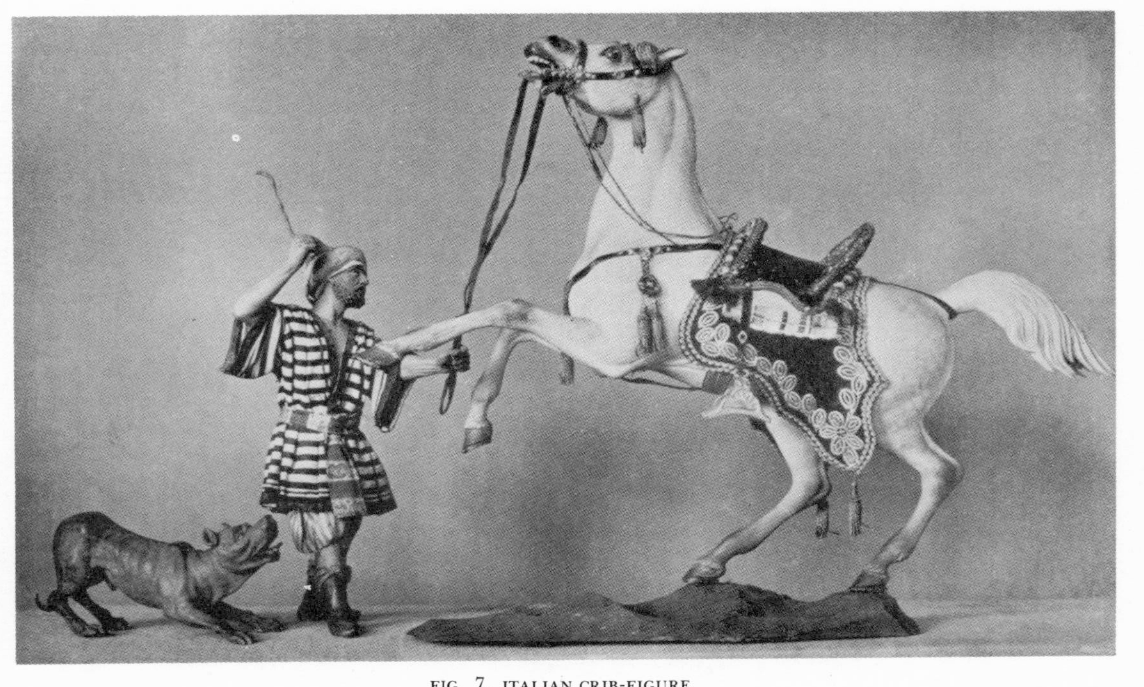

FIG. 7. ITALIAN CRIB-FIGURE
From *Denkmäler der Krippenkunst*, by R. Berliner (B. Filser, Augsburg)
Bayerisches National-Museum, Munich

there to stimulate their imagination. Legend declared that at the time of Christ's birth an annual fair was being held at Bethlehem; there was not a being between heaven and earth who was not present at it. There were peasants, shopkeepers,

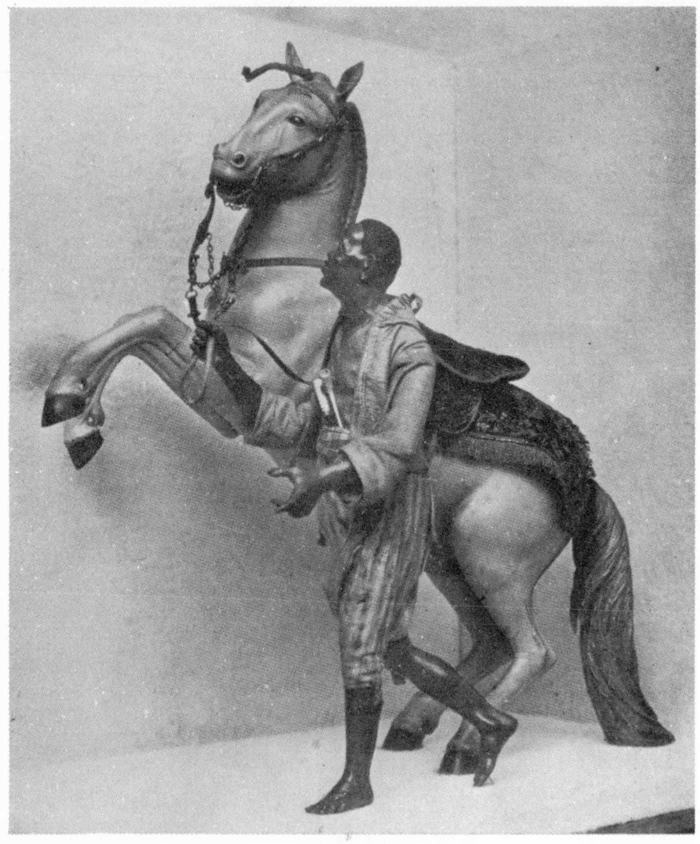

FIG. 8. ITALIAN CRIB-FIGURE
Naples
Bayerisches National-Museum, Munich

workmen, landowners, merchants, beggars, cutpurses, fighting, dancing, playing, music-making children, domestic and exotic animals, accompanied by an improbable wealth of *finimenti*. Here, too, were all sorts of food—vegetables, fish, bread, fruit, cheese, sausages, eggs, oysters, lobsters, macaroni, etc.—besides musical instruments, gold and silver plate, and so on. Still further opportunities were offered by the arrival of the three kings and their magnificent retinue. All these figures were transplanted to the eighteenth century and to the Gulf of Naples, so that these cribs can be regarded as mirrors of the contemporary life of the people. The illusion of time and place was the greater because the artists liked to make use of local scenery and to present, with all the illusion of perspective,

FIGS. 9, 10. ITALIAN CRIB-FIGURES, EIGHTEENTH CENTURY
The procession of the three kings. In the original the two sections here shown are continuous.
Bayerisches National-Museum, Munich

its ruins, peasant houses, inns, brooks, waterfalls, and bridges. Automata were also sometimes employed, and travellers who saw the Christmas cribs in Calabria refer to them as theatres of moving puppets.

The Christmas crib was a sport in Naples to which all, high and low, from royalty down, paid homage. Charles III liked to make his cribs with his own hands, and his queen, a Saxon princess by birth, made the dresses for her husband's dolls. Families were in the habit of visiting each other's cribs, some of which, it is said, cost as much as 30,000 ducats. All those who visited Naples at this period—Goethe, the Abbé de Saint-Non, Gorani, Friederike Brun, etc.—are full of praise for them. "The artistry displayed completely defies description and passes the bounds of imagination."

Whereas the Neapolitan cribs were concerned in particular with the exceedingly vivid folk scenes, the multitude of figures completely distracting attention from the main incident—the birth of Christ—the Sicilian cribs preferred the gruesome. On

FIG. 11. ITALIAN CRIB-FIGURES, EIGHTEENTH CENTURY
From *Denkmäler der Krippenkunst*, by R. Berliner (B. Filser, Augsburg)
Bayerisches National-Museum, Munich

that account they liked to bring within the scope of their subject-matter the massacre of the innocents as well as the Nativity. In wealth of captivating detail and art of arrangement the Southern Italian cribs of the eighteenth century are unsurpassed, but the art was not neglected in the rest of Italy, being especially encouraged in different centres by the Franciscans and the Capuchins. In 1709 a rich prelate in Rome had a whole houseful of crib-figures, the entire collection being valued at about 8000–9000 thalers. Among the cribs in Roman churches those of S. Francesco a Ripa and S. Maria Ara Cœli were deemed to be the most lovely. The latter was built round the famous *Bambino* and, influenced by local tradition, included the Emperor Augustus and the sibyls among its figures.

FIG. 12. ITALIAN CRIB-FIGURES, EIGHTEENTH CENTURY

Group of shoemakers. From *Denkmäler der Krippenkunst*, by R. Berliner (B. Filser, Augsburg)

Bayerisches National-Museum Munich

In æsthetic beauty and in richness the German cribs cannot be compared with the Italian; but maybe the German examples have one advantage, in that they were dear to the heart of the people. The dramatic instinct of simple folk as expressed in the cribs succeeded in creating a wholly unliterary but, in the best sense of the word, popular art. It sprang from the hands of the folk themselves, and hence was able, without the necessity of any explanation or interpretation, to count on the full understanding of the crowd. In order to display the wealth of figures on a wider scale subsidiary events were appropriated from Biblical history, the crib being so divided into five groups: the Nativity, the shepherds' offering, the visit of the three kings, the flight into Egypt, and the Holy Family in Nazareth. Such scenes as the massacre of the innocents and the marriage at Cana could be added *ad libitum*. Legends too could be introduced there to the heart's content, and since the cribs were usually on display from Christmas to Candlemas the ambition of the artists was fully satisfied, and to those artists belonged every man who could use a wood-carving tool and every woman who could sew and embroider.

In old Bavaria, the Tirol, Upper Austria, and Styria peasants and hunters moulded and carved during the long winter evenings, year in and year out, for a genuine crib was never really finished—it could always be made still more beautiful and brought nearer to perfection. There are, indeed, cribs which took a hundred years and more to complete. In the Tirol not only did every church possess its crib, but in many villages every house had its own example. In the towns too the cribs had their amateurs who, if they could not carve themselves, would buy the figures from others. In Augsburg crib-figures were already articles of commerce in the eighteenth century, while in Munich there was even a regular crib market. A priest at Volkmannsdorf, near Moosburg, in 1678 ordered from an itinerant wood-carver a crib that took two years to finish. The figures, 30–35 cm. high, were wooden dolls with detachable limbs, movable heads, glass eyes, and wigs. In Munich the cribs of the Michaelshofkirche, with their figures 1 m. high, enjoyed great fame, and similar renown attached to those of St Peter's church and the church of the Franciscans.

In the convents these cribs were a welcome diversion, giving the nuns employment the whole year round. One of the most valuable belonged to the Ursuline convent at Innsbruck. Its oldest figures dated from the beginning of the eighteenth century; these were 20 cm. high and made of wood, with legs and arms

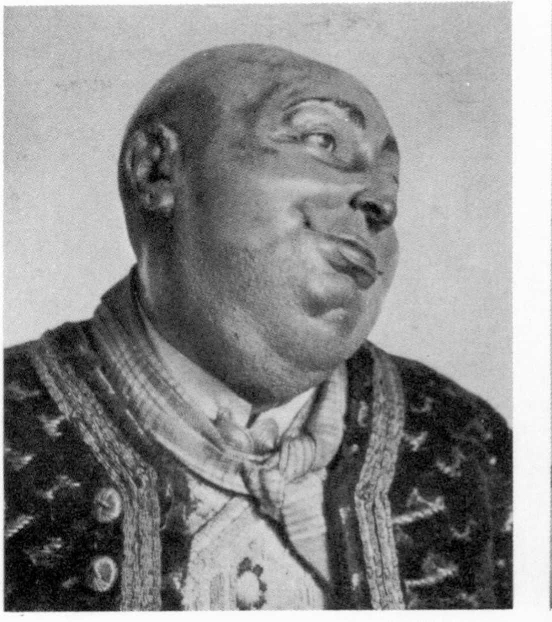

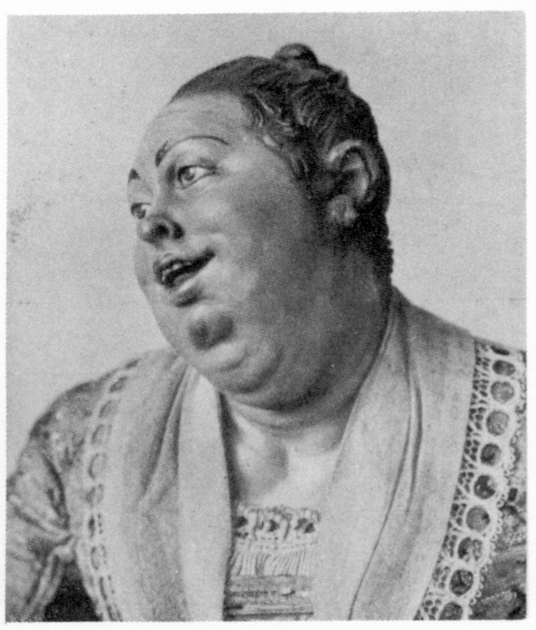

FIGS. 13, 14. HEADS OF ITALIAN CRIB-FIGURES

Host and hostess From *Denkmäler der Krippenkunst*, by R. Berliner (B. Filser, Augsburg)

Bayerisches National-Museum, Munich

of the same material rendered movable, and thus adjustable to any position, by means of wire joints. The heads were moulded

FIG. 15. ALPINE WOODEN DOLL
Eighteenth century. German crib-figure
Bayerisches National-Museum, Munich

in wax, with hair of flax or wool; the bodies were wrapped round tightly with thick linen, and then dressed in real cloth garments of an exceedingly rich sort in the rococo theatrical style. The angels thus appeared with short hooped dresses and laced sleeves, like the *ballerinas* of the time; they wore velvet and silk trimmed with ermine, embroidered with gold and silver,

FIG. 16. WOODEN DOLL: PEASANT BOY
Eighteenth century. German crib-figure
Bayerisches National-Museum, Munich

ornamented with beads and stones, some with crowns containing real jewels. The retinue of the three kings boasted the uniforms of the queen's guards. Armour, weapons, and accoutrements were of metal—some even of precious metal. A private individual, even at that time, could not easily rival such a display, but the wood-carvers had the power at least of providing infinite variety in their figures. There were the three kings, with their retinue of servants, horses, camels, and elephants, besides shepherds, hunters, wagon-drivers, butchers, woodmen, hermits, beggars, men and women of the peasantry, cows, sheep, chamois, hares, etc.—all carved with care and either painted or dressed. In

FIG. 17. CRIB-FIGURES OF LIME-WOOD DRESSED IN CLOTH
Tirol. Eighteenth century

one of the most famous cribs—that at the church of Birgitz—the figures were so expensively clothed that a mantle of one of the kings cost thirty-six florins.

The accumulation of architectural detail is as characteristic of the Tirol cribs as the superabundance of figures. The crib belonging to the Jaufental family, of Wilten, which has been placed in the Österreichische Museum für Volkskunde, in Vienna, mingled dwellings of the Tirol Alps with architecture of the Renaissance, temples, and a hill castle. For the six scenes it represented it utilized twenty-four buildings, 256 human and 154 animal figures. Its origin is to be traced to the year 1700. As automata were by that time widely known, these were eagerly seized upon for the purpose of animating the figures. Paul von Stetten saw such automatic cribs at Augsburg in 1779. When

FIG. 18. CRIB OF THE STIFTSKIRCHE IN ADMONT
J. Thaddäus Stammel. Painted by Pötschnick, 1755

thus equipped they might cost anything up to several thousand guilders. Canon W. Pailler in the reminiscences of his youth describes a large mechanical crib which he saw in Upper Austria; all trades were represented there—smiths, cabinet-makers, tanners, millers, threshers, turners, shoe-makers, tailors, carpenters, etc., engaged in their respective activities.

The pleasure taken in the representation of popular life wholly surpassed that taken in the sacred episodes. There is a saying in Upper Bavaria, describing a general hotchpotch, which runs: "There's as much hash here as in a crib." For this reason an episcopal ordinance at Regensburg in 1789 prohibited the display of popular accessories in the cribs, and others banned them completely. Thus, Count Thürheim in 1803 forbade the exhibition of the crib in Bamberg, on the ground that the inhabitants of the Frankish provinces were too enlightened for such frivolities. When the reaction set in against the preceding rationalism there was a brief renaissance of the crib in the nineteenth century, but this was only a waning gleam, a last flicker before coarse-grained materialism overwhelmed or destroyed all the natural springs of folk life.

Yet still were there artists at work whose charming creations were able to hold the public interest in the crib and to provide it with ever new life. Roman Anton Boos, a distinguished sculptor, was a maker of crib-figures, and above all there was that Ludwig who died in 1830 and who, in Hager's judgment, produced the best figures ever made by Munich crib-makers. His figures, 26 cm. high, with movable heads and jointed limbs, were delicately and carefully carved to show muscles, veins, and wrinkles, like the late Gothic wood statues. Even then large sums were given for them by amateurs. The Court musician Zink in 1838 paid forty-one florins for three shepherd figures, and the crib belonging to Heindl, director of the mint, was valued in the sixties of the century at 1500 florins. About this time, however, interest in the crib began seriously to diminish. With Andreas Barsam, who died in 1869, disappeared the last artistic carver of crib-figures at Munich, and Sebastian Habenschaden, who died in 1868, was their last skilful painter.

The crib maintained itself longest, as an object of real popular affection, in the Alpine Tirol and in the Salzkammergut, where wood-carving as a home occupation is pursued from sheer love of the thing itself. So long as the carvers took the baroque statuary as their model their figures retained the lifelike movements and the rich clothing of that style; the first indication of a decline is marked about the year 1800 in the work of Johann

Mühlmann, who looked for his model to contemporary religious painting, with its anæmic and academic rigidity. The *naïveté* of the conception declines, for, although the display of figures is certainly not less rich, these are no longer so free from restraint as formerly, their deportment being now clearly determined by the episode at the manger. In place of individualized figures appear types of a wholly general sort; the fantastic baroque

FIG. 19. CRIB-FIGURES OF LIME-WOOD DRESSED IN CLOTH BY JOHANN KIENINGER IN HALLSTATT
Nineteenth century

landscape which predominated in the eighteenth century takes on a simpler form, which is nearer akin to reality, and often preference is given to a background of one simple tone. As the foreground of the general panorama was rendered plastically and the background merely painted, great care was taken to secure an illusive transition from one to the other.

The Tirol, with its wood-carved and painted figures, remains the true home of the crib, and has produced in many homely workers artists in their own kind. Such an one was discovered at Hallstatt in the person of Johann Kieninger, who died in 1899, at the age of seventy. The Österreichische Museum für Volkskunde, in Vienna, possesses a mountain crib made by him, which, in Haberlandt's words, is

> a wholly charming work, in which the mixture of popular tradition and individual inventive spirit is worthy of special note. Unsur-

passable and inimitable is the admixture of traditional and personal conception in this work, which impresses the observer with a sense of childish innocence and of native homeliness.

An artist of similar powers was Josef Partsch, in Engelberg, who died in 1886, at the age of seventy-five. His speciality was the Christmas crib with multitudes of participant figures. Praise has been given to his work for its touching simplicity in spite of its religious profundity and its great technical skill. Partsch tinted his figures in water-colour; the smaller ones were cast in plaster from stone moulds.

In the Salzburg district dressed figures were preferred, for they were, according to Mühlmann, far more national. The heads were hand-moulded of flesh-coloured wax, with eyes of glass, the bodies themselves being concealed under the stiffly extended folds of cloth. These clothes, with their trimmings of gold spangles, beads, and precious stones, made a rich show. Composition in groups, a popular method in the eighteenth century, was here more rarely employed, each figure standing by itself, without any direct relation to its neighbour. The most magnificent crib in the Tirol style is that splendid one of the master tanner Moser in Bozen, which is now in the possession of the Bayerische National-Museum; this was made in the forties by its former owner and by Johann Pendel in Meran. Here a richly fantastic architectural setting is equipped with all kinds of mechanical devices, such as clocks and waterfalls, while the staffage includes several hundred little figures varied in size from 2 to 10 cm. in accordance with the rules of perspective. This work of art cost its owner 10,000 florins. In the Tirol the wood-carvers had been in the habit of wandering about at Christmas with their works—Annunciation cribs, the three kings, and Easter plays—exhibiting them for a trifle, but about the year 1900 this custom came to an end, and the crib sank into being a museum piece.

The Riedinger Collection, in Augsburg, containing many hundred figures, was deservedly famous. Larger and more varied was the collection of Max Schmederer, particularly rich in Southern Italian cribs. These were offered by the owner to the Bayerische National-Museum, in Munich, but the then director of the museum rejected this valuable gift, and only a Press campaign, inaugurated by the Munich *Neueste Nachrichten*, succeeded in getting this collection, which is unique in its own way, accepted and put on exhibition. Since that time it has formed one of the most noteworthy sections of the museum.

Art and science first paid attention to the crib when it had already ceased to play a part in the life of the people; since then the cribs have been collected by museums and regarded as valuable documents for a knowledge of the folk. Friends of the crib in all quarters have banded together, and out of their confederation a working society has been formed. In Munich the architect de Crignis conducts a crib school every winter, and P. Simon Reiter, of the Order of St Francis, has written a practical textbook on the subject. The Salzburg Crib Society (*Krippenverein*) in 1919 organized a large exhibition of ancient cribs in St Peter's Monastery, while in the winter of 1928 an exhibition of cribs, arranged by the German Catholic Women's League, was held at Berlin, a place which we usually associate with crude materialism. Alongside Oberammergau cribs from the carving workshops of the well-known portrayer of Christ, Anton Lang, were to be seen works of the progenitor of the German crib art, Sebastian Osterrieder, of Munich. The Innsbruck sculptor Kuen showed a wood-carved crib panorama; Knapstein, of Cologne, a Rhine citizen crib; the wood-carving school of Warmbrunn a crib by Professor Del Antonio; and the art school of Münster several works by the director of their sculpture class, Professor Guntermann, in which the group of Mary and the child captured attention by its divine beauty. The wood-carved crib-figures of Melchior Grosseck in Silesian popular dress made a lively and naïve impression; deeply rooted in Westphalian sentiment were the homely lines of Mormann's figures. A large Christmas group with artistically carved jointed figures by Lamers-Vordermayer was on loan from a private collection, as were also a Westphalian crib from a convent of Poor Clares and a beautifully cut wood crib frieze. Works of Tirolese, Ukrainians, and Berliners stood here side by side. Majolica cribs and wooden cribs made by machinery (among them one made after Dürer) were to be seen here beside reliefs and religious scenes executed in needlework. These had as great an interest for their reflection of folk customs as for that of folk dress, which no organization in the world can now revive.

The crib with automatic figures approaches close to the puppet theatres. Such cribs with moving figures were once so common in Aachen that the expression *Krepche* ('crib') simply signified a puppet-show. In Poland too boys used to tramp round with the *Schopa*, a box with a crib, in front of which danced various kinds of dolls. In the second half of the eighteenth century the puppet-crib-plays were a favourite amusement for high and low in Vienna, Schönholz styling them the "most beloved miniature theatre with thumbnail actors, all movable." Maria Theresa patronized Frau Godl, whose real name was Barbara Müller, and who had set up her stage in Lerchenfeld. Her principal scenes were those of the three kings, the flight into Egypt, the massacre of the innocents, Holofernes' tent, and Saul's palace, while the puppet theatre At the Metal Tower (Beim Blechernen Turm) in Wieden presented with special care Joseph's flight into Egypt, Daniel in the lions' den, and David and Goliath. In Vienna only scenes from the Old Testament were allowed; when the painter Sacchetti, in 1806, desired to present Christ's Passion by means of wax puppets permission was denied him.

It seems that in France preference was always given to cribs with automatic figures. The Theatines in Paris even in the seventeenth century were accustomed to set up a Christmas crib with movable wax figures at the door of their cloister, and as late as the eighteenth century the crib scene and the Passion were represented and played with similar puppets on the Petit Pont de l'Hôtel Dieu. This kind of crib-puppet-show was disseminated throughout the whole of France, gaining special popularity in the south at Marseilles. The cribs in Lyons too must have been more like marionette theatres than the motionless groups such as appeared in Italy and Germany. It is assumed that they were a relic of the old mystery-plays, with which indeed they share one peculiarity—the introduction into their text and personnel of comic elements expressed in lower-class dialect. In Lyons these were represented by Father and Mother Coquard, who mingled with the shepherds at the crib and spoke in the Lyonese *patois*. They sang a duet in which the mists of their native place were referred to, and which ended with an exhortation to the youthful spectators to behave themselves well. Thus in France, as in Germany, the crib, by quick transition, turned into the puppet-show.

IV

THE TIN SOLDIER

So far as we have referred hitherto to the toy doll we have been concerned (whether directly expressed or not) with the toys of the little girl. About 90 per cent. of toy dolls are of the female sex, and they are treated as playthings almost exclusively by girls. James Sully has asserted that boys are not generally inclined to play with dolls, and, if they do turn to dolls, they

FIG. 20. LITTLE HORSEMAN
Terra-cotta. Cyprus

FIG. 21. LITTLE HORSEMAN
Terra-cotta. Tanagra

(so far, at least, as the boyhood of the United States is concerned) like only such as are out of the ordinary—clowns, for instance, negroes, and Eskimos—and prefer even animal toys to these. Evidence of this can be found even in the distant past. It is recorded from early medieval Iceland that the six-year-old Arngrim Thorgrimsson presented his little brass horse to his four-year-old uncle Steinolf Arnorsson since he himself was then too old to be still playing with it.

The boyish toys of past centuries were mostly dolls representing warriors and horsemen, large numbers of which, made of wood, clay, and metal, have been preserved from various epochs of civilization. Even in late classical days little Trojan horses with warriors on them, whether as toys or mementoes of a journey, were sold on the site of ancient Ilium. Innumerable are the extant figures of little knights made of baked clay or glazed stoneware, the playthings of German boys of the fifteenth and sixteenth centuries. The earliest pictorial record we possess of the existence of such German toy dolls is a miniature from the well-known Codex of the Herrad von Landsperg, which in 1870 was destroyed by fire in the Strasbourg Library. In this late twelfth-century manuscript a couple of boys were shown making two knights, held on horizontal strings, fight with one another. This "knightly toy," which made tenderest childhood

FIG. 22. LUDUS MONSTRORUM
About 1160. A children's toy or a puppet-play?
Painting from the Codex of the Herrad von Landsperg

acquainted with its future duties, lost, in the course of centuries, nothing of its popularity. Emperor Maximilian, the "last of the knights," who had his life written and illustrated in the *Weisskunig*, commissioned Burgkmair in 1516 to execute a picture which shows how as a boy he used to play with a friend at tilting. They are seen pushing two mounted knights in armour set on wheeled frames one against the other, each endeavouring to unhorse the other with his lance. The Emperor took delight in this toy even in his old age, for he ordered from the armourer Koloman, helmet-maker in Augsburg, two such knights in so-called tilt harness set on wooden horses. They were intended for the young King Ludwig II of Hungary. Several similar toys have been preserved—for example, in the Museum Ferdinandeum at Innsbruck, at Burg Kreuzenstein, and elsewhere. In the Vienna Kunsthistorische Hofmuseum there is a toy of this kind certainly earlier than the sixteenth century, representing chargers and knights of cast brass ready for the tourney. These can be drawn toward each other with strings so that their tilting lances meet. The finest example, however, is to be found in the Bayerische National-Museum, in Munich. This must belong to about the year 1556 and bears the arms of the Nürnberg family Holzschuher. The horse's neck and feet are movable; the knight

FIG. 23. EMPEROR MAXIMILIAN, WHEN A BOY, AT PLAY
Woodcut by Hans Burgkmair. From the *Weisskunig*

is made of wood and is fully accoutred. Armour and weapons are technically correct down to the minutest detail, so that the doll may be regarded as a model of contemporary harness for man and charger. Less expensive must have been the "little wooden man which when pulled can fight" with which about this time the six-year-old Felix Platter was presented as a *Dockenhansl* ('gift'). Even as late as 1600 the old Duke William of Bavaria was accustomed to give princely persons "little tilting horsemen which were moved by clockwork."

The warrior doll has, it is true, remained a toy beloved of boys, but it has been considerably simplified in recent times. The more or less complicated mechanism of the plastic figure has given place to the flat figure, represented in the tin soldier. Originally the flat figure was made of lead, but as that substance

FIG. 24. MAN ON HORSEBACK WITH THE HOLZSCHUHER CREST
About 1556
Bayerisches National-Museum, Munich

was too soft and had too little power of resistance it was replaced first by tin and then by an alloy of tin and antimony. Robert Forrer and Theodor Hampe have traced back through centuries the genealogical tree of this little flat figure, and have rummaged out for it a very respectable ancestry—in which they count not only the so-called *schardana* figures of the close of the third century B.C., but the primitive metal warrior dolls found in Etruria, Greece, and Istria. In the Hallstatt period, about the year 1000 B.C., the records increase. At Karnten, near Rossegg, a primitive little horseman was discovered in a grave-mound; even thus early it was made of a mixture of lead and tin. In some graves near Frögg eighteen similar horsemen were found. The famous bronze chariot of Strettwag, in Steiermark, which belongs to the later Hallstatt period and the significance of which even now has not been quite explained, seems to be full of such dolls. The horsemen, measured with their shields, are 13 cm. high, the foot-soldiers 10½ cm. Very similar figures were known in the Chibcha civilization. From the lake of Siecha, Cundinamarca, Columbia, was fished up a golden group representing a raft with a high priest and his retinue of ten persons. The little figures, conventionally flat, like tin soldiers, measure 3 to 7 cm. in height.

FIG. 25. ROMAN TIN FIGURE FOUND ON THE RHINE
Legionary of the Imperial period
British Museum

The Romans were also acquainted with flat soldiers made of tin and lead. In a grave at Pesaro was discovered a figure of Cæsar on horseback, and at Mainz was found a one-sided cast

FIG. 26. LEAD SOLDIERS
Fourteenth century

tin figure of a Roman legionary of the Imperial epoch, with short sword, large shield, greaves, and helmet. This art was not lost in the Middle Ages, but it is not at all clear whether the little figures which have come down to us from that time were devotional objects, the so-called 'pilgrims' badges,' or toys. Perhaps they may have been both. Little cast figures of lead and tin belonging to the thirteenth and fourteenth centuries have been found in the Seine; these, about 6 cm. high, represent horsemen such as St George and St Martin, and are finished only on one side.

Theodor Hampe mentions a Schlüsselfeld table decoration of 1503, a famous goldsmith's work of Nürnberg provenance, which represents a ship made animate by the presence of many little figures. These were made of precious metal and enamelled; they show, at any rate, that men knew at that time how to make toy figures of the tiniest form. Whether they were made then in cheaper material to serve as toys is not known. It is said that Louis XIII in 1610 played with lead soldiers which were 7 cm. high. These lacked foot-plates and must consequently have been stuck on to the table. In 1650 an army of soldiers was made of silver for the twelve-year-old Louis XIV, Georges Chassel, of Nancy, sketching the designs, which were carried out by the goldsmith Merlin. The cost of this royal toy came to 50,000 thalers. When the King and Colbert a few years later wanted a number of toy soldiers for the Dauphin they sent to Nürnberg, as the most famous manufacturing centre of toys of all kinds and of inventions connected therewith. At the King's order the goldsmith Johann Jakob Wolrab made several hundred silver cavalry soldiers and infantry. It is believed that the famous Vauban was sent to Nürnberg expressly for the purpose of supervising the work. The 3½-inch figures were equipped by the compass-maker Hans Hautsch and his son Gottfried with an automatic device which aroused great admiration. In 1698, about twenty-five years after it was completed, Weigel writes of it from memory:

> They went through the usual war manœuvres very ably; they marched to left and to right, doubled their ranks, lowered their weapons, struck fire, shot off, and retreated. Then the lance-men tried to knock the cavalrymen out of their saddles, but these were quite prepared to defend themselves by firing their pistols.

Another army, of cardboard figures, consisting of twenty squadrons of cavalry and ten battalions of infantry, was made by Pierre Couturier, called Montargis, between 1670 and 1671 and by Henri Gessay between 1669 and 1670. It cost 28,963 francs, and is still to be seen in the public collections in Paris.

The tin soldier was not a common thing in the seventeenth century. When in 1670–75 the Prince Elector of Bavaria desired a military toy for his son, who later won fame in war as Maximilian Emmanuel, he ordered the carver Matthias Schütz to make him some wooden infantry and cavalry. *Papier mâché* must also have been in use for this purpose, for the father of the well-known painter of hunting pictures Johann Elias Ridinger made some soldier dolls at Ulm of that material. The tin soldier is a creation of the eighteenth century, and owes its being, like the contemporary porcelain plastic, to Germany. All authors are in

FIG. 27. OFFICER AND TROOPERS OF THE LÜTZOW CORPS
Painted tin figures of the eighteenth century
Germanisches National-Museum, Nürnberg

agreement that it arose as an "echo of the victories of Frederick the Great." At any rate the mass production of these figures begins with the Seven Years War. From these cultural efforts the Fatherland gained at least one benefit in that the lead-soldier industry throughout the nineteenth century remained specifically German. By the year 1900 there were about twenty German tin-soldier factories with an annual output of a million marks. "Germany," remarks the well-known French social historian Henri d'Allemagne, "in particular has improved this toy. Skilfully she has taken advantage of its influence on the upbringing of the child—to set alight and to nourish the flame of patriotism and to keep alive the traditions of honour and bravery."

With the toy soldier too Nürnberg asserted its old preeminence as the centre of the toy industry. Andreas Hilpert, of Coburg, in 1760 settled in Nürnberg, and was able to make this toy popular by means of his technical and artistic ability. He invented the small flat figures with a standing plate such as we have to-day, and created in the diverse types a genuine art form—"these most artistic tiny rococo figures," as Hampe calls them. About forty different kinds of Prussian, French, Russian, and Turkish soldiers came from his skilled hands; his activities, however, were not confined entirely to military figures, but included also representatives of civil life. After the artist's

FIG. 28. WOOD-CARVED FIGURES OF SOLDIERS
Eighteenth to nineteenth centuries
Germanisches National-Museum, Nürnberg

death the firm was carried on successfully by his family. In the illustrated catalogues of the great toy-shops of that time, that of P. F. Catel at Berlin in 1790, and that of Georg H. Bestelmeier at Nürnberg, issued between 1798 and 1807, the tin soldiers are fully represented. At the same time the picture sheets intended to be cut out became popular. About the year 1810 a baker named Boersch at Strasbourg cut out of paper all the regiments of Napoleon's Great Army, painting them carefully and correctly. He was said to have compiled in this way about five thousand figures. These sheets were once very common and were often designed by well-known artists. War-painters, like Adam, were, of course, in special demand for this kind of work, but the picture sheets of Pettenkofen, Schindler, Kriehuber, Zampis, and others are also noteworthy for their boldness of treatment.

In the first half of the nineteenth century the Nürnberg industry found open competitors in Northern Germany. The firm of G. Söhlke was founded at Berlin in 1819, and this name for long generations remained dear to the children of that city.

FIG. 29. TIN SOLDIERS
About 1820

This firm produced, in addition to soldiers, figures for fairy-tales, for *Gulliver's Travels*, and for *Robinson Crusoe*; keeping fully abreast of the time, it manufactured toy trains when as yet the railway had hardly gained its footing, and accommodated itself, too, to the success in fiction of *Uncle Tom's Cabin*. In 1830 J. E. Dubois founded his business at Hanover, producing figures of the Hanoverian legion, the Napoleonic guards, etc., magnificently executed in design, modelling, and painting. In spite of all this, æsthetic and commercial precedence remained with Nürnberg, for Ernst Heinrichsen, who set up business in 1839, outstripped all competitors. He it was who introduced figures of standard sizes, fixing these at 3 cm. for infantry and 4 cm. for cavalry; formerly the figures with muskets had been 11 cm. high and the Prussian guards 8 to 9 cm. He himself was a very skilful designer and engraver, and he was able to attract famous artists to work for him; he got Camphausen, for example, to sketch for him Gustavus Adolphus's Swedish troops. In addition to this, he followed the tendencies and interests of his time with a keen eye to commercial opportunities. He prepared tin figures for Cooper's Indian tales, for polar expeditions, and for African explorations besides others for the Trojan War, the Crusades, and bull-fights; the militia of 1848, the Crimean War, the Boer War, the Great War, and, lastly, the Reichswehr have all been represented accurately by this old-established firm, for it is still in existence. In addition to Heinrichsen must be mentioned Johann C. Allgeyer, of Fürth, whose firm has been carried on by his son and grandson, and J. C. Haselbach, of Berlin, who made over 5000 different moulds for figures representing soldiers in the English wars in India, the French wars in Mexico, and the war of 1870–71.

Playing with soldiers is by no means confined to boys; it has been popular also with those monarchs who wished to advance militarism among their subjects. Certainly we may not count among these Tsar Peter III, the first of the Holstein family to sit on the Russian throne, for, although this half-witted prince liked to play with toys, he did it in a wholly childish way; once, indeed, he ordered a rat which had eaten two of his soldiers of *Tragant* to be tried and sentenced to death according to martial law. The tin soldiers which he played with are still preserved in one of the small palaces of Oranienbaum. In the style of the tin soldiers was the war game invented by the Polish general Mieroslawski and practised seriously both at headquarters and among the rank and file, although it was not played with figures, but with little stones. Tsar Nicholas I was said to have excelled in this game and to have beaten even his brother-in-law, who later became Emperor William I, at the "battle near Bautzen." Hampe declares that for this Tsar Heinrichsen made tin soldiers representing the mounted regiments of the Russian guard.

The Prussian monarchs, on whom the spirit of militarism was so deeply impressed, have left in their model dolls valuable evidence relating to the old Prussian army. Of these the arsenal and the Hohenzollern Museum in Schloss Monbijou possess extensive collections, and to them Adolf Menzel has devoted a careful examination. The oldest piece is a great red-and-black-striped wooden doll belonging to the first years of the reign of Frederick William I. It represents a grenadier of the red life-guards about to throw a grenade. Then come models in very much smaller style, for the most part representing officers and men of the Prussian army in the second half of the eighteenth century, very delicately and accurately formed, with wax heads and cloth hats, bearing real plumes. Frederick William III possessed dolls of this kind 20 cm. high, moulded in lead, and then painted in perfect likeness of the uniforms and accoutrements worn in all the regiments of his army. Emperor William I too had a doll army, each a careful model of a soldier from one of his regiments. The Armee Museum in Munich has a similar collection of little toy soldiers representing the former Bavarian army.

What originally was only a plaything for half-grown boys—no less a person than Goethe refers to it in his *Dichtung und Wahrheit*—has gradually become, thanks to careful modelling and equipment, an aid to students of military history and a valuable museum piece. The Germanische National-Museum, in Nürnberg, made a start in the collection of these objects, and its example has been followed by various other local museums. In the Landes-museum at Dresden whole battlefields are reproduced by means of tin soldiers. The Great War much stimulated interest in this representation of reality. A Berlin factory in 1915 had window displays in which the most famous historical battles, beginning with those of the Egyptians and concluding with the battle of the Masurian Lakes, were reproduced by the employment of 35,000 tin soldiers. In the same year an exhibition of 50,000 tin soldiers was held in Vienna at the KK. Österreichische-Museum, in the Stubenring; art students from Professor Breitner's school undertook the preparation of the various landscapes necessary. The battles of Custozza, Sedan, and Mars-la-Tour were shown by the utilization of some 8000 figures.

To-day the tin soldier has almost wholly lost its position as a toy; it is now preserved by the collector in the vitrines. These collectors are so numerous and so zealous in their activities that they have formed societies, and have established special periodicals devoted to their sport. The tin-soldier museum of the aulic councillor Anton Klamrot in Leipzig is justly famed: "German antiquarians and students of folk life truly owe much to him because of his energetic activities in the collection of these objects; in this he has shown real historic sense."

V

THE PUPPET-SHOW IN ANTIQUITY

When we turn to the history of the marionette theatre we are often, as noted already, in doubt whether we are dealing with the hand puppets or with the marionettes on strings. Inferences, however, may be drawn when the references to the marionettes occur in metaphors where they are introduced as objects of comparison. When Aristotle writes that those who direct the

FIG. 30. ACTOR (MESSENGER) AND COMIC ACTOR (DANGER)
Terra-cotta. From Myrina

marionettes need only pull their strings in order to set in motion first the head and hands of the little being, then its eyes, shoulders, and limbs, all so delightfully obedient, it is quite clear what sort of puppets he alludes to. Apuleius describes the strings in the same way. Galen likens them to men's muscles, and Plato relates them to our passions, which pull us this way and that. Horace compares in his satires (*Duceris ut nervis alienis mobile lignum*) the human lack of free will with the stringed marionette.

Puppets with movable limbs have been preserved from ancient times in considerable numbers, but genuine marionettes controlled by strings are certainly not among them—not to mention puppets with movable eyes. Perchance they have all perished, for the material of which they were made—wood—is especially liable to decay; this is particularly unfortunate, for we know that the puppet theatres were very popular in Greece. Xenophon, in describing a visit he paid in 422 B.C. to the house of Kallias in Athens, refers to a Syracusian who came with his puppet theatre to entertain the guests, but who could not capture Socrates' attention. In the time of Sophocles the marionette theatres must have appeared frequently in Athens. In his

FIG. 31. WINGED DOLLS

Terra-cotta. From Megaris and Tanagra

Deipnosophists Athenæos reproached the inhabitants of that city because they had handed over the theatre of Dionysos to the marionettes of the *neuropastes* ('string-puller') Potheimos, and because they took more delight in these than in Euripides' plays. The Emperor Marcus Aurelius likewise interested himself in these puppets; he, like Horace, makes comparison between their strings and man's free will. This comparison, however, is so obvious that no special ingenuity is demanded for its invention. The Indian *Mahabharata* also mentions the string-controlled marionettes, and compares their servile condition with human beings. Rajah Sekhara, writing at the beginning of the tenth century A.D., makes two movable puppets take part in one of his dramas. The name given to the puppet-showman was *sutradhara*, which means literally 'string-puller,' and this name eventually passed over to be applied in general to the theatrical producer —a proof that puppet-plays, which even to-day still form the only dramatic entertainment of Indian rural communities, must be more ancient than the theatre of human actors.

The Fathers of the Church did not fail to connect their cheap moral reflections on mankind with the mechanism of a well-made puppet. Considering, indeed, the great antipathy they displayed toward all manifestations of heathen culture, including the theatre, it is astonishing how tolerantly they dealt with the marionette. Clement of Alexandria, Tertullian, Synesius, all of whom condemned the theatre, said nothing against the puppets, which, we must assume, could not have been so obscene as, for example, the mime.

FIG. 32. WINGED EROS
Clay doll from Myrina

How and when the puppet-play came to Germany we do not know; perhaps we may agree with Philipp Leibrecht in assuming that the introduction of this art was due to the jugglers who followed the Roman legions over the Alps. Evidence in favour of this theory is provided by the fact that an Old High German gloss identifies *Tocha* (the modern *Docke*, or doll) with *mima*. Its existence, however, cannot be definitely proved before the twelfth century, and even then proof is forthcoming only if we take the picture in the Codex of the Herrad von Landsperg, which belongs to about the year 1170, not as a children's toy, as it seems to be, but as a puppet theatre. Immediately after this date, on the other hand, records begin to multiply, demonstrating clearly that this thing must by then have become universally popular. In 1253 the minnesinger Meister Sigeher compared the way in which Pope Innocent IV behaved toward the German princes with a puppet-play—a comparison which unfortunately was only too fully justified. Ulrich von dem Türlin, Willehalm von Oranse, Thomasin von Zirclaria, writing in the thirteenth century, agree in styling "mundane joys a mere puppet-play." Hugo von Trinberg, who finished his great didactic poem *Der Renner* in the year 1300, relates that the jugglers used to bring small puppets from under their cloaks, making the spectators laugh with their antics. They were called by various names—*Kobold*, *Wichtel*, and *Tatermann*. The laughter occasioned by their jests gave rise to a proverb—'laughing like a *Kobold*.' Apparently these puppets—and here one must think rather of hand puppets than of stringed marionettes—contributed much to strengthening the popular belief in pigmies and little imps. In the Redentin Easter play which a Cistercian, Peter Calf, composed in 1464 at the village of Redentin, near Wismar, Luzifer speaks of those "who play with puppets and cheat fools of their money." In the German translation of the French heroic poem *Melagys*, which belongs to the fifteenth century, a scene lacking in the original is inserted wherein the Fée Oriande performs a play with two puppets—in all probability hand puppets.

FIG. 33. GREEK CLAY DOLL FROM A GRAVE
Marionette?
Biardot Collection

The oldest picture, moreover, which we possess of the puppet theatre on this side of the Alps shows figures of this kind worked by hand. These are, as might be supposed, figures in the head of which the performer places his forefinger, while he moves the arms with his thumb and middle finger. Only three-quarters of the puppet accordingly is seen, and, since usually but one performer with two hands at his disposal operates them, only two puppets appear at one time. In the manuscript of a French heroic poem, *Li Romans du Bon Roi Alixandre*, written in 1338 and illustrated with miniatures in 1344 by the painter Jehan de Grise (a native of Flanders?), are to be found two pictures of a puppet-show. In one of the little pictures three maidens watch a performance which some scholars have attempted to connect with Punch and Judy; in the other four little male spectators watch a scene of slashing and stabbing. In these oldest illustrations is to be observed an element which has remained a property of the hand-puppet stage up to the present day, preserving its force of attraction from of old—the motive of quarrelling and conflict and cudgelling. This codex is in the Bodleian Library, Oxford.

The general popularity of the puppet theatre is proved too by the attitude of popular writers. Luther, for example, once called the papacy "a public puppet-show," and speaks elsewhere of the "holy puppets." Direct references, however, in illustrative form and text are in reality extremely few, but for this there is a double reason. The itinerant jugglers with their puppet-shows did not belong to any "respectable trade," and therefore scholars, in whose hands rested the literature of the day, were ashamed to confess that they had found any pleasure in the activities of a vagabond class of society. They held it beneath their dignity to take notice of jugglers and puppet-showmen, and if on occasion a comparison involving these did slip from their pens, as in Luther's case, it was only incidental.

FIG. 34. GREEK CLAY FIGURE FROM A GRAVE
Marionette?
Biardot Collection

We know that puppet theatres certainly existed before the sixteenth century, but it is impossible to tell how they were constituted or of what sort were their puppets and repertoire. All documentary evidence, unfortunately, is silent on these points. We learn that Count Jan von Blois in 1363 ordered a puppet-show to be given in Dordrecht, that in 1395 a man was paid for such shows, that he had presented a puppet-play before the Count of Holland, that in 1451 a ban was laid on puppet-shows during Easter, but beyond that the records do not go. What plays were given then and what sort of puppets were employed we cannot tell. No doubt Rabe is right in assuming that they were hand puppets exhibited in a kind of Punch-and-Judy show.

Considering the miserable conditions of the wandering entertainers whose bread and butter depended on these shows, we may well assume that all the stage arrangements were as primitive as could be. As regards the nature of the repertoire, however, only conjectures of a wholly general sort are possible.

All those authors who have occupied themselves with this question assume with Gustav Freytag that fighting and buffeting must have come first in the programme. This is the more

FIG. 35. NIKE IN TERRA-COTTA
Marionette? From Myrina

probable in that performances of the kind still hold their charm for the common people. The buffoon was certainly there; he could not be banished even from the serious mysteries.

These comic characters are similar to one another in all countries. The Indian Vidusaka, the Greek *mimos*, the Javanese Semar, the Turkish Karagöz, the German serving-lad Rubin, from whom Hanswurst and later Kasperle derived, all belong to one family. One and all they are ugly, dirty, vain, impertinent, greedy, cowardly, and coarse; their chief activity lies in giving sound cudgellings, in return for which they receive, if possible, sounder beatings. About the end of the fifteenth century and the beginning of the sixteenth the expression *Himmelreich* ('kingdom of heaven') was introduced into Germany for the puppet-show (or perhaps only for the box of the puppet-showman). It is to be found, for example, in Thomas Murner's

FIG. 36. MEDIEVAL PUPPET THEATRE

Miniature by Jehan de Grise, from a manuscript, apparently Flemish, *Li Romans du Bon Roi Alixandre*. Painted between 1338 and 1344

Bodleian Library

Die Narrenbeschwörung and in various Nürnberg civic degrees of the fifteenth and sixteenth centuries, and is transferred even to the players in the form of *Himmelreicher*. No plausible explanation of this word has so far been offered. T. Hampe believes that it is not impossible that the title was conferred on them because of their repertoire, which commonly derived its material from Biblical sources, known by the showmen to be popular among their audiences. Thus, a certain Heinrich von Burgund in 1510 wished to produce in Nürnberg his puppet-show of Christ's Passion, but the town council refused to give their permission. Leibrecht, working on old German civic decrees of the second half of the sixteenth century, has published a whole series of records relating to performances of religious puppet-plays in Nürnberg, Lüneberg, Nordlingen, Augsburg, Danzig, and Berlin; these amply demonstrate the great popularity of this kind of show. Not one of the texts is now known; no doubt they were handed down by oral tradition. There was only one sixteenth-century German author who wrote for the puppet stage, and that was Hans Sachs. The artistic level of the pieces of the Nürnberg shoemaker lies fairly low, and we must assume that the performances of the puppet-showmen of that time raised still slighter claims to literary worth, the less indeed in that a good many of them must have been improvised. In point of fact there is no real evidence that these puppets were introduced with dialogue at all; some historians, such as Magnin and Maindron, are of the opinion that the puppets appeared only in pantomime, and that a man in front of the stage related the course of action.

VI

MARIONETTES IN THE SIXTEENTH TO THE EIGHTEENTH CENTURY

In the sixteenth century there developed out of the puppet type the special form of the marionette. Our knowledge of its existence at this period comes from a man who was then widely famous as a scientist—Hieronymus Cardanus, an Italian, whose book *De Varietate Rerum* appeared in 1557 at Nürnberg. He writes there:

> I have seen two Sicilians who did real wonders with two wooden figures which they made to move. A single string was carried through both. It was attached on one side to a fixed post and on the other to the leg which the showman moved. The string was stretched at both sides. There is no kind of dance which these figures could not execute. They made the most astonishing movements with their feet, legs, arms, and head—all with such varied gestures that I am unable, I confess, to render an account of such an ingenious mechanism, for there were not several strings, some tight, some loose; only one went through the figures, and that was always tight. I have seen a good many other wooden figures which were set in movement by several strings sometimes tight and sometimes loosened; that is no marvel. I must add, too, that it was a really pleasant thing to see how the gestures and steps of these puppets synchronized with the music.

In *De Subtilitate Rerum* the author returns to this subject, remarking:

> If I were to relate all the wonders which are carried out by those stringed puppets popularly called *magatelli* a whole day could not suffice for me, for they play, fight, hunt, dance, blow trumpets, and attend artfully to preparing their meals.

The puppets which the scholar describes with such ardour belong to the kind which the French call *marionettes à la planchette*. In Italian they were called also *fantoccini*, whence comes the French *fantoches*. They were known everywhere, especially in Italy, where they seem to have been invented and whence wandering entertainers took them beyond the Alps to display their activities at annual fairs. The showman, who made the puppets dance by moving his limb, accompanied their steps on some kind of instrument—flute, tambourine, or bag-pipes. They were even taken across the Channel to England, Hogarth

FIG. 37. ITALIAN MARIONETTES "À LA PLANCHETTE"
Etching by J. Dumont. 1739

introducing them in the print which represents the annual fair at Southwark in 1733. The German traveller Adam Olearius, who in 1633 made a long journey by Russia to Persia, met them too in the farthest eastern parts of Europe.

In the sixteenth century the puppets supported on and moved by strings or wires received the since commonly accepted name of 'marionette.' In Italy, where they originated, they were given various names—*fantoccini*, *burattini*, *pupazzi*, *bambocchie*—but it is impossible to determine what distinctions these titles indicated.

Seemingly the puppet-showmen came from Italy to France with the actors of the *commedia dell'arte*. For many years Catharine de' Medici, herself a Florentine princess, played, as the Queen Mother, the chief *rôle* at the French Court and patronized

FIG. 38. MARIONETTES "A LA PLANCHETTE"
Anonymous French engraving. About 1800

Italian fashions and entertainments. The name 'marionette' is to be discovered first in a work of Guillaume Bouchet, *Les Serées*, about the year 1600. Its etymology is uncertain. Magnin is inclined to regard 'marionette' as a diminutive of 'Maria,' since small statues of the Virgin have sometimes been so called, but how the movable and, for the most part, comic puppets could have received their name from a motionless object of devotion he does not explain. In the second edition of his excellent book the same scholar conjectures that it may be derived from 'Marion,' the pet-name of the heroine in one of the Robin and Marion pastorals. These dance songs of the twelfth century, however, must have long been forgotten by the sixteenth century, and could hardly have afforded the occasion for the invention of a character name. Magnin also notes the similarity

"Les Petites Marionettes"
From *Goût du Jour*, Paris, about 1810

PUNCH-AND-JUDY SHOW IN FLORENCE
Lasinio. About 1780

of the word to *marotte*, the fool's sceptre or baton, while Frisch in his *Lexicon* of 1741 refers to the connexion of the word with the medieval fool names *morio*, *morione*. The etymology, as has been stated, is still not absolutely settled; all we can say is that about the end of the sixteenth century and the beginning of the

FIG. 39. MARIONETTES "A LA PLANCHETTE"
Lithograph after a sketch by Carle Vernet. About 1820

seventeenth the name is in regular use and accepted everywhere. In Germany the movable puppet was called *Dattermann* or, as Konrad von Haslau gives it, *Tatermann*, a title which has now sunk into entire oblivion.

In Italy during the sixteenth century attempts were made to perfect the marionettes. Federigo Commandini, of Urbino, who died in 1575, and Giovanni Torriani, of Cremona, who accompanied Emperor Charles V into a monastery, are mentioned as improvers of the mechanism, but it seems that their efforts were directed principally toward the perfection of automata. Their creations, however, were so startling that Torriani was immediately suspected in Spain of being a magician, while in France public accusations were made against certain wizards who possessed little devils, called marionettes, to which they offered sacrifice and from which they sought counsel. This testifies, at any rate, to the excellent quality of the puppets, which, we must suppose, appeared extraordinarily lifelike.

FIG. 40. MARIONETTE THEATRE
Engraving of the eighteenth century
Germanisches National-Museum, Nürnberg

In 1573 the first permanent theatre of Italian marionettes was established in London; this made an impression on Shakespeare, for he mentions these puppets repeatedly, and on one occasion makes Hamlet wish to be the speaker for a marionette stage. Many, indeed, were the supporters of the puppets. Ben Jonson in his *Cynthia's Revels* makes a woman say: "As a country-gentlewoman, [I should] keep a good house and come up in term to see motions." In 1609 a new puppet theatre, in which French marionettes were shown, was opened in London. They must have been of great perfection, for Ben Jonson has occasion to refer to their movable eyes. The first English puppet-showman known by name is Captain Pod in 1599. In England these marionette theatres seem to have aimed directly at strong stage effects; at least in the second half of the sixteenth century the accusation is made that they aimed too much at deluding the eyes of the spectators. They introduced plays dealing with such subjects as the fall of Sodom and Gomorrah, the destruction of Jerusalem, and the burning of Nineveh, in addition to much later subjects, bringing thus on their boards the murdering of the Guise brothers, the Gunpowder Plot, etc. During the performances a showman spoke in front of the curtain, explaining the actions on the stage—a method which is the same as that employed in the Spanish marionette theatres. Thus, in the second part of *Don Quixote* Cervantes describes a marionette performance where the *titero*, being behind the scenes, moves the strings unseen by the audience, while a boy in front relates the events. The puppets, he adds, were so charming that the pitiable knight was deceived and came to aid the unfortunate princess against the insolent Moors.

FIG. 41. WITCH
Movable. Of lime-wood dressed in cloth. Eighteenth century
From a peasant theatre in Steiermark

In the permanent puppet theatres of the larger Spanish towns, in order to create a greater illusion by keeping the showmen out of sight, the custom began of providing the puppets with a dialogue of their own—a development which makes its appearance also in England. To alter the tones of his voice, and so differentiate the various characters, this showman made use of a small instrument which he put in his mouth when interpreting certain parts. It was made like a *cri-cri*, of two pieces of metal, tortoise-shell, or ivory tied together with strips of cloth. In speaking it was held between the tongue and the roof of the mouth, and the showmen were always in danger of swallowing it. In Spain it was called *pito*, in France *sifflet-pratique* or simply *pratique*, and in Italy *fischio* or *pivetta*. The Italians must have used it for various parts; in Germany, however, according to Rabe, it was reserved for devils alone.

Although there is no doubt that in the second half of the sixteenth century puppet theatres were established in France, the first reliable records concerning showmen known to be actively engaged at certain places and at definite times come from the second half of the seventeenth century; these records relate to a family named Brioché (Jean, Charles, and François), who were apparently Italians by birth and whose original patronymic was Briocci. They gave their shows on the Pont Neuf, in Paris, and had the honour in 1669 to be summoned to Saint-Germain for the purpose of entertaining the Dauphin. The Briochés have always been referred to by French authors as marionette-players, but it seems, however, that they worked with hand puppets, an opinion which Rabe shares. The fact that a masked ape, Fagotin, took part in their shows indicates the latter, for such a performer would only have caused confusion among string puppets. This famous beast was stabbed to death by Cyrano de Bergerac, but it had many successors; all French puppet-showmen of that time got monkeys to aid in their performances, and they were all named Fagotin. Later on a cat was introduced into the French Guignol in the same way as the Hamburg Kasperle theatre introduced a dove, the English Punch a dog, and the Viennese Wurstl a rabbit. One of the Brioché family was so skilled in his art that during a tour in Switzerland he was suspected at Solothurn of being a wizard, and was able to escape with his life only by an immediate flight. Indubitable string marionettes were those with which the director La Grille produced his operas from 1676 in the Marais district. In the year following his house was named the Théâtre des Bamboches, and attracted spectators because of its richness in costume and stage setting. Its machinery permitted him to give even fairy operas with ballets and scenic changes. The marionettes in Paris had many adversaries, among whom were to be counted not only the clergy, including the famous Bishop Bossuet, but actors of the regular stage as well. At one time the little puppets are putting the latter to ridicule, pitilessly making them a laughing-stock, the marionettes always having the laughter on their side; and then, during the last years of the seventeenth century, the actors, thinking that their takings were being reduced by the competition of the puppet theatre, proceed to drive back the puppet stages to the markets in the Parisian suburbs.

FIG. 42. OLD WOMAN
Marionette. Modern

At that time Germany was in a very turbulent state, and it need hardly be said that the wandering puppet-showmen, under the weight of circumstances, must have suffered greatly. They had also to face competition, on the one hand from the touring actors, who, as, for instance, the well-known director Weltheim,

FIG. 43. DON QUIXOTE DESTROYS THE MARIONETTE THEATRE
Engraving in a French edition of *Don Quixote*. Eighteenth century

played alternately with living persons and puppets, and on the other from the itinerant charlatans and quack doctors who sought to attract the public by all means in their power, making use even of the puppet stage. Between 1625 and 1659 the orphan asylum at Hamburg received a taxed contribution from the 'comedian-doctor' in the hop market. A comedy of the creation of the world was shown with puppets in 1644 at the Saxon Electoral Court in Dresden and in 1646 at the same Court in Moritzburg. In 1656 an entertainer gave performances in the market-place at Hamburg, showing how the King of Sweden was shot by the Poles and was carried off to hell. The puppet theatre of Michael Daniel Treu seems to have been exceptionally fine. When he visited Lüneburg in 1666 his repertoire consisted of twenty-five pieces, including *King Lear*, *Titus Andronicus*, and *Doctor Faust* as well as historical pieces such as *General Wahlstein*, and *Cromwell's Ghost*. From 1681 to 1685 he gave performances at Munich. At this time Leibrecht says that the town of Basle was the centre of marionette-showmen from all the chief countries of Europe, their appearance there giving a decidedly international atmosphere to the town. About the end of the seventeenth century a certain Johann Hilverding stood out predominant above the mass of puppet-showmen; from the year 1685 he worked in Vienna, made considerable tours, got as far as Stockholm, and then, once more returned to Vienna, associated himself with F. A. Stranitzky, a low-class comedian then held in great esteem. He gloried in showing over fifty comedies and operas with his figures one and a half ells high—"the figures execute all *actiones* like living persons with fitting movements."

The string puppets predominated in the sixteenth century; in the seventeenth century the hand puppets, along with their typical figures Polichinelle and Punch, took precedence. Magnin assumes that Polichinelle, in spite of his close kinship with the Italian Pulcinella, was a French creation; he may be right, for this figure, in its arrogance, its impudence, and its boastfulness, provides a brilliant synthesis of the French character. His partner was Dame Gigogne, a comic figure indigenous to the folk stage, who represented the small tradespeople. About 1650 the pair had already a permanent theatre at the Porte de Nesle, in Paris.

The first reliable German records relating to this character date from the second half of the century, and once more it was foreigners who were concerned with the entertainments. In 1649 a hydrologist, Manfredi, made his appearance in Nürnberg; apparently he was "the first to exhibit Polizinell with small puppets." In 1657 Petro Gimonde, an Italian puppet-showman, introduced the type at Frankfort. The wood-carver Matthias Schütz, besides the various toys which he made between 1670 and 1675 for the Bavarian princes, "carved also various wonderful heads for a *Meister Hämmerlein* show, and also two hands therefor which were hollow within so that one might control them by this means with the fingers of each hand." This is the earliest detailed record we possess relative to the German hand

FIG. 44. FRENCH MARIONETTE THEATRE
End of the eighteenth century. Contemporary engraving

puppets of this period, and for the first time, too, this record gives them a special name. In 1744 a lengthier definition is provided by Frisch—"*Meister Hämmerlein*: the kind of puppet-show

FIG. 45. FRENCH KASPERLE THEATRE
From an engraving. About 1820

where the hand is put inside the puppet, thereby moving the body, head, and arms; it is the *Pickelhäring* ['pickle herring']." This puppet seems generally "to have had an ugly masked face."

FIG. 46. KASPERLE THEATRE IN PARIS
About 1825

Pickelhäring is a comic figure, also called Hanswurst and, in the next century, Kasperle; the last-mentioned provided the whole show with the name by which it is still known.

The most famous member of this family is the youngest of the line, the English Punch. His name is simply Pulcinella or

Punchinello adapted to the English tongue, though he seems to have taken a roundabout journey to England instead of coming straight from Italy. Whether the Stuarts brought him back with them from France at the Restoration or William of Orange introduced him from Holland is not certain; at any rate, he first appeared after the Revolution of 1688, to become, with his

FIG. 47. PORTABLE MARIONETTE THEATRE
French lithograph. About 1830

dog Toby, a favourite of the English people. In the course of years his character must have changed, for in 1697 Addison could still describe him as a jolly, somewhat blustering petticoat-hunter in French style.

In the eighteenth century the puppet theatre played a considerable *rôle* in the public life of all civilized countries—a fact amply proved by the severe attacks it met with from those who belonged to the regular stage. If they could damage it and restrain its annoying competition they did so; and if they could not succeed in that they at least slandered it and sought to discredit it in contemporary opinion. How the Parisian regular stage gave performances alongside the puppet theatres has already been seen, and the same was true of England. There the incessant attacks of the Puritans at last succeeded in getting the theatres closed, but the marionettes were overlooked or considered harmless, and so were allowed to continue their performances—whence great indignation. About the year 1577 Geoffrey Fenton had written a book to demonstrate that the puppet-showmen were as unworthy creatures as the regular actors. The actors, for their part, complained loudly because the marionettes were specially favoured. In 1642 the company of Drury Lane complained of the special privilege shown toward the puppet stage and demanded that it should be closed. In eighteenth-century France too the actors did all they could to render the showmen's life difficult. These were not permitted, for example, to produce their puppets with dialogue—only monologues were allowed, and even then not spoken in a natural voice, but distorted by means of the *sifflet-pratique*.

Their affairs in Germany were no better. To the ranks of their enemies naturally belonged all those concerned with the opera and the regular theatres, such as the Hamburg barrister Barthold Feind, to whom the opera of his native town is so deeply indebted, Count Seeau, the director of the Court theatre at Munich, who refused to permit the puppet theatre to play at the annual fairs, and, toward the end of the century, the historian of the Hamburg theatre, Schütz, who calls them "miseries," dangerous and demoralizing. "The pitiful trick puppets worked by vagabonds with thumb and forefinger"—that is to say, the Kasperle theatre—is attacked by him with special hatred. The vulgarity and obscenity in which the showmen indulged alarmed the clergy. When Sebastian di Scio played *Faust* at Berlin in the first years of the century his success was so great that Philipp Jakob Spener asked for an order of prohibition from the authorities, on the ground that he considered it dangerous for the devil to be conjured up in this piece and made to appear on the stage. The many enemies of the marionette stage had a temporary success, at any rate in Prussia. On June 3, 1794, the provincial authorities were instructed to drive off all unlicensed marionette-showmen, for "these unworthy vagabonds were seeking to win applause by indecent innuendoes."

Nevertheless the puppet stage proved itself competent to deal with all attacks made on it. It remained popular, since, in Germany, as in England and France, everything the regular theatre excluded or, at least, tried to exclude—as, for example, the German Hanswurst—took refuge on this convenient stage.

The puppet-showmen as a class became so numerous that in the eighteenth century they were formed into a kind of guild with their own special regulations and customs; one peculiar rule was that none of the play texts was to be written out, all, including the prompter's stage directions, having to be learned by heart. As a theatrical director wore a red vest to distinguish him from the rest of the troupe, so the puppet-showmen adopted a characteristic attire, consisting of a large black cloak and a broad-brimmed hat. Many of the principal showmen gave performances alternately with human actors and with puppets, which seemingly were fairly large (Matthison notes that the marionettes at Strasbourg were half life-size); automata too were introduced, and surprise turns were given special attention. On one puppet stage the chief attraction was a soldier with a pipe in his mouth who puffed out the smoke. Reibehand, who gave performances in Hamburg from 1728 to 1752, introduced in *The Prodigal Son* a hanged man who fell in pieces from the gallows, put himself together again, and then pursued the hero. It was he who got up the *Öffentliche Enthauptung des Fräulein Dorothea* (*The Public Beheading of Miss Dorothea*); after the execution applause broke out, so the head was put back on the puppet, and it was decapitated once more. In the rich commercial towns, such as Frankfort, the puppet theatres were brilliantly fitted up and equipped with all sorts of technical devices, the finest of these being that of Robertus Schaffer at the Liebfrauenberg, which was patronized even by high society. The Margrave of Baden-Durlach had Court players in his service for his marionettes; in 1731 at Berlin their director, Titus Maas, gave a performance of *The History of Prince Menschikoff*, a highly realistic theme. In his Hungarian castle at Eisenstadt Prince Nicholas Esterhazy established a puppet stage the richness of which was in keeping with its owner's wealth. For this Pietro Travaglia made the artistically formed figures, and between 1773 and 1780 Joseph Haydn wrote for them five operettas: *Philemon and Baucis*, *Genoveva*, *Dido*, *La Vengeance accomplie*, and *La Maison brulée*. These were presented in pantomime on the stage, with the singers behind the scenes. Esterhazy's puppet theatre won such fame that Maria Theresa requested a special performance to be given by it at Schönbrunn.

Haydn himself had a small marionette theatre and also wrote the operetta of *Le Diable boiteux* for Bernardone's puppet stage at the Kärntner Tor, in Vienna. At Mannheim in 1767 the officers of the Palatine regiments established a marionette theatre, the dolls for which were made by the Court sculptor

Egel. It was opened with Molière's *Don Juan*; as in all the Court theatres of that time entry was gratuitous. In Vienna and Hamburg the puppet stages seem to have been more beloved than anywhere else, but even in Augsburg, which he visited in 1781, Friedrich Nicolai found at a marionette performance "much better society than he had expected."

We must not forget what a great influence the marionettes have had upon the poets. In the Christmas of 1753 Goethe

FIG. 48. GOETHE'S PUPPET THEATRE
Goethe Museum, Frankfort

and his sister received from their grandmother a puppet theatre which he enthusiastically recalls in *William Meisters Wanderjahre* and in *Dichtung und Wahrheit.* He himself wrote for the puppets *Das Jahrmarktfest zu Plundersweilen*, and in his first years at Weimar he was still working for them. Above all, however, "the honour is unquestionably due to the wooden theatre of the Frankfort marionettes for having sowed in the soul of the young artist the seeds which later flourished into the most important poem in German." Assuredly Leibrecht is in the right when he says that this does not cast honour on the contemporary puppet-show, for Goethe's *Faust* stands above the puppet-comedies just as the moon stands above the pond in which its light is reflected. Yet *Faust* was and has remained for the following century the piece by which the puppet-showmen have gathered their largest audiences. In Vienna it was presented, with inserted *arias* and ballets, oftener than *Don Juan*, which was *Faust's* only rival. Every puppet-showman had his own *Faust* which he

FIGURE OF KASPERLE, EIGHTEENTH CENTURY
From Karl Gröber's *Kinderspielzeug* (Deutscher Kunstverlag, Berlin)
Landesmuseum, Linz

PUNCH-AND-JUDY SHOW IN NAPLES, 1828
Mörner

himself had adapted to meet his own requirements. Scheible in the fifth volume of his *Kloster* has published five different texts which had been written for Ulm, Augsburg, Strasbourg, and Cologne. For the success of these pieces we know the reason, which was that the chief *rôle* was taken by Hanswurst. Hanswurst, who became Kasperle at Vienna in the beginning of the eighteenth century, was always the best figure among the

FIG. 49. FAUST AND KASPERLE
Puppet-figures by Guido Bonneschky. About 1840

marionettes. He moved his head, arms, feet, and hands; he rolled his eyes; he could open his mouth and stick out his tongue. He improvised as much as he wanted, and offered the public what they delighted in most—cudgelling, vulgarity, and still again vulgarity—as much as their hearts could desire.

Magnin regards the first quarter of the eighteenth century as the most brilliant period for the marionettes in France. Men of spirit and wit, such as Lesage and Piron, wrote jolly pieces for the puppet theatres in which they indulged in mischievous sallies at the regular stage. The actor and poet Favart, who later became famous for his vaudevilles, first made his appearance in 1732 with marionettes which had the right of entry to Court, since the Duchesse du Maine, a daughter of Louis XIV, had introduced them in the entertainments she arranged at the Château de Sceaux. In one piece she had ridicule cast on the Académie; this presented Polichinelle pleading his right to be elected one of the forty immortals. The Duchesse du Maine and the Duchesse de Berry also produced farces at Versailles in which the victories of Marshal Villars were rendered ridiculous, a peculiar expression of admiration for a successful commander-in-chief. The marionettes conquered the *salons* of high society when the cardboard figures, the *pantins* mentioned already, came into fashion, and puppet-plays were improvised everywhere. The marionettes even intruded into Cirey, otherwise consecrated to purely philosophical speculations, and led Voltaire himself, as Mme de Graffigny writes in 1738, to associate himself with them and write verses for them.

From the fifties of the century public interest in the marionettes declined. Whether the players failed in fitting wit and humour, or whether preference was given to the surprising effects of mechanical devices, at any rate a change was sought for. About 1740–50 the puppet stage was imitating the much-admired scenic work of the painter engaged at the Paris Opéra, Servandoni, but the shadow theatre, newly introduced through the rococo passion for everything Chinese, supplanted for a time the simpler charm of the familiar marionettes. In the second half of the eighteenth century the Parisian puppet theatres moved over from the suburbs and from the markets to the boulevards so as to be nearer the public threatening to abandon them, and thus able to remind it of their presence. The terrible years of the Revolution did not spare them, harmless though they were. Camille Desmoulins relates that scenes were presented on the puppet stage in which Polichinelle ended his life under the guillotine; the revolutionaries, however, went even further, for in 1794 a married couple were beheaded only because their Polichinelle was considered too aristocratic and had ridiculed *Père Duchesne*—the *Rote Fahne* of the period.

In this epoch the hand-puppet theatre developed in France a distinct characteristic type of its own—Guignol—although that originated not in Paris, but in Lyons. Its inventor was Laurent Mourguet, who gave it its name, the etymology of which is uncertain. Mourguet provided Guignol with the costume of a silk-worker, and made him speak in the dialect of the common people. His hero is ignorant, but acute of ear; unscrupulous, but ready to assist; good-hearted, always in good humour, sceptical to the last degree, but when he is flattered easily led by the nose. His partner was at first Polichinelle until Mourguet invented the *rôle* of Gnafron, a character rich in not very refined jests. With the aid of a friend to whom he gave the *scenarios* Mourguet, who died in Vienna in 1844 at the age of ninety-nine, himself wrote all the pieces he played. They were considered witty, and possessed to a high degree the flavour of Lyonese life—a fact which guaranteed their continuous success.

In England the marionettes gained the right of entry to Court under Charles II. From Pepys' diary we learn that on October 8, 1662, they performed at Whitehall. On August 30, 1667,

FIG. 50. ENGLISH PUPPET THEATRE
Designed by B. Pollock, with a scene, *The Silver Palace*. About 1860
Presented to the Victoria and Albert Museum by Mrs Gabrielle Enthoven

Pepys visited a fair and to his astonishment met Lady Castlemaine, the King's mistress, in a puppet theatre where the play of *Patient Grizell* was being given. Themes from legend and the Bible remained popular in England throughout the whole of the next century, but the puppets also gave dances of jigs, sarabands, and quadrilles. One of the most eminent puppet-showmen in England was that Powell whom Addison and Steele popularized in their weekly papers. He appeared alternately in London and in Bath during the season. His strength lay in burlesque of the Italian opera, a dramatic form undoubtedly challenging satire. The dramas of Shakespeare too were adopted by the puppet stage. Samuel Johnson, whose judgment was so deeply esteemed by contemporaries, found that the marionettes played much better than living actors, giving it as his opinion that *Macbeth* was much more impressive in the puppet theatre than on the regular stage.

The eighteenth century was the period when Punch flourished in England; at this time he was the representative of English folk humour with all its degeneracy and peculiar qualities. Punch must very soon—certainly by 1731, when he is described just as he is to-day—have assumed the appearance by which he is recognized now—the bird-like features, with the huge nose, the hunches in front and at the back (for physical deformities always amuse the mob, who find afflictions of this kind ridiculous), and lastly his peaked cap and the ruff round his neck. By 1713 he had already a permanent theatre in Covent Garden—a proof of his popularity. Swift with good reason attributes this popularity to the effrontery and shamelessness of his *rôle*, but a type which receives as well as gives so many cudgellings is sure of applause. Just as no German play, however serious its theme might be, could be counted complete without its Hanswurst, so in England Punch had to appear even in Biblical dramas if these were to prove popular. Thus, in 1703 he played a part in a puppet-play *The Creation of the World* and in 1709 in a comedy *The Flood*. Yet there was a difference between the two characters. Hanswurst is a dull wag who thinks he is witty when he wallows in the mud of vulgarities and double meanings, while Punch out-devils the devil. To-day when Punch is named we think of the famous journal which bears his name, but we must not associate its polite tone with the old hand puppet. The latter's speech and action were of an unparalleled brutality; Magnin draws a comparison between the Punch of that time and Henry VIII. Punch openly speaks of atrocities, and revels in things repulsive and disgusting. He delights in paradoxes, and expresses his opinion without respect to propriety and good manners; the word 'shame' does not enter into his vocabulary. At the end of the century the puppet company was enriched by the appearance of Judy, Punch's wife, who from that time on remained a permanent member of the English troupe of hand puppets.

If civilized nations regarded the puppet stage only as a secondary kind of entertainment, the marionette theatre formed for a long time the only dramatic art among nations without their own culture, as, for example, the Czechs. In the eighteenth century Josef Winizky and Matthias Kopeckj were famous puppet-showmen among them. They were forced to write their plays themselves, owing to the fact that none except the uneducated populace understood the Bohemian tongue.

VII

MARIONETTES IN THE NINETEENTH CENTURY

THE early years of the nineteenth century saw German literature in the grip of a romantic sentiment which brought to the puppet theatre a still greater interest, if that were possible, than Goethe had with his attachment to classic standards. The spokesmen of the new movement, who were also responsible for the new science of Germanistics, regarded everything from a genuinely national point of view. Arnim and Brentano collected the folk-songs which still lived among the peasantry, and hoped to be able to draw materials from the puppet repertoires suitable for their purposes. They believed that the puppet theatres had a connexion with the old mysteries, and thought of the treasures which might be discovered there. These, however, did not materialize, and those things which the young poets themselves composed for the marionettes did not get very far. Like the puppet-shows given in the houses of Brentano, Achim von Arnim, etc., these were unknown outside the narrow circle of those who participated in the performances. The greatest talent for this *genre* was possessed perhaps by E. T. A. Hoffmann, whose simple humour was peculiarly adapted for the marionette style, which always hovers on the border-lines of humanity. Yet, although he occupied himself much with them in his house, he has left nothing of this kind behind him. He possessed, as Ölenschläger records, a cupboard full of marionettes which he loved to manipulate, and derived great amusement from startling his guests with them.

In various *feuilletons* there has been mentioned the "brilliant study" which Heinrich von Kleist dedicated to the marionette theatre. This reference, however, shows only that the writers had heard but half the story, and that they could not have read the article in question. This appeared under the title *Concerning the Marionette Theatre* on December 12–15, 1810, in a Berlin evening paper, but on account of the severe theatre censorship it was written purposely in a very obscure and cautious style, and has really nothing to do with the puppet theatre. The poet starts with the thesis that man, who

FIG. 51. KASPERLE THEATRE
Nineteenth century

had eaten from the tree of knowledge and had thereby lost his instinctive natural life, must return through ever broader,

FIG. 52. FAIR BOOTH, WITH NÜRNBERG PUPPETS AND KASPERLE
Woodcut by A. L. Richter. About 1850

higher, divinely striving knowledge and self-resignation to his lost innocence, regaining paradise after his wandering through the world. At all intermediate positions in this development man remains imperfect; therefore in the theatre as well only the marionette or the god is perfect. From the standpoint gained by these philosophic-æsthetic inquiries Kleist criticizes the Berlin actors and dancers. Had this article not borne so famous a name it would long ago have disappeared from the marionette literature.

The interest of the professional authors of that time is shown by the facts, for example, that August Mahlmann in 1806 published a volume of small satirical dramas for the marionettes and that the Berlin author Julius von Voss also wrote for them. Up to that time the items in the repertoire had been handed down only by oral tradition, and even then they adhered closely to certain themes, the religious pieces such as *Adam and Eve*, *David and Goliath*, *The Prodigal Son*, and *Herod* remaining in vogue up to 1840. Schütz and Dreher, whose stage was set up in Berlin and Potsdam, gave performances, in the old style, of *Faust*, *Don Juan*, ancient legends like *Genoveva*, and biblical stories like *Esther and Haman*, *Judith and Holofernes*, and *The Prodigal Son*. In the first quarter of the century the puppet theatre of Geisselbrecht was regarded as the finest in Germany. This man, a mechanic by trade, had come from Vienna and performed all over Germany. It was he who served as model for the puppet-showman in Theodor Storm's novel *Päle Poppenspäler*. Countess Line Egloffstein wrote on December 14, 1809, from Weimar to her sister Julie about his entertainments.

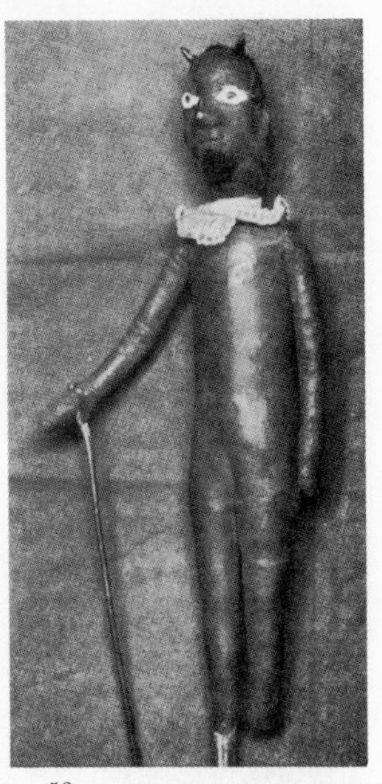

FIG. 53. ANCIENT FIGURE OF A DEVIL FROM WINTER'S COLOGNE PUPPET THEATRE
From *Das Rheinische Puppenspiel*, by Carl Niessen

> Yesterday at last we saw our charming marionettes, and were so bewitched that we got quite foolish about them. Scenic changes and decorations were worthy of the greatest admiration, and the little Arlequin was so lovable he deserved to be kissed. He eats and drinks like a man, smokes his pipe, loves his Colombine, and charms every one who comes his way.

The writer of this letter betrays in her description what the public enjoyed—the lifelike appearance of the puppets. Geisselbrecht's marionettes could even move their eyes, cough, and spit! These ingeniously constructed figures were augmented by trick puppets, or metamorphoses, invented by Franz Genesius,

FIG. 54. HÄNNESCHEN STAGE OF THE OLDEST STYLE
From *Das Rheinische Puppenspiel*, by Carl Niessen

FIG. 55. PEASANTS OF THE OLD COLOGNE HÄNNESCHEN THEATRE

which, by means of a certain mechanism, could be made to change their appearance; a girl dancer was thus turned into a balloon, a pumpkin into a man dancer, and a mushroom into a dwarf. Geisselbrecht's figures, too, could discharge flint-lock guns, draw their swords, pour out wine, etc.

The marionettes provided each audience with what it desired. At Berlin in 1851 104 consecutive performances were given of a silly farce by Silvius Landsberger, *Don Carlos, the Infanta of*

FIG. 56. NOTABLES OF THE OLD COLOGNE HÄNNESCHEN THEATRE
From *Das Rheinische Puppenspiel*, by Carl Niessen

FIG. 57. THE PEASANTS OF THE COLOGNE PUPPET-PLAY
Figures from the theatre of H. Königsfeld, junior. After old models
From *Das Rheinische Puppenspiel*, by Carl Niessen

Spain, while the marionette stage of Weyermann at Ulm was on such a high level that the famous historian of that city, Professor Hassler, called it the "National Theatre of Ulm." The director realized where success was to be gained; he attired his puppets in

Ulm dress and made them speak in the Ulm dialect—types of such a kind as were calculated to move, delight, and inspire his audiences. Cologne owes its most original creation to the director Christoph Winter, who founded the Hänneschen Theatre in 1802, the continuous success of which is directly traceable to its strongly marked local characteristics. The inventor of the type,

FIG. 58. HÄNNESCHEN IN THE WEHRGASSE
Painting by Passavanti
From *Das Rheinische Puppenspiel*, by Carl Niessen

who died in 1872, at ninety-six years of age, possessed a peculiar talent for making situations and persons ridiculous without giving offence, and for creating a comic scene without having recourse to vulgarity. His types included Hänneschen, Mariezebill, Neighbour Tunnes, introducing in a jovial but harmless manner institutions of the town of Cologne and casting a satirical light thereon. He graduated his expressions according to the audience he had before him, dividing these into three classes—children, adults, and Sunday visitors. To the last-mentioned he spoke in the roughest terms. All his plays had to end happily

—even the most tragic of dramas, such as *Romeo and Juliet*, which closed with the marriage of the lovers. An attempt made by Millowitsch to transfer the Hänneschen Theatre to the regular stage met with disaster.

FIG. 59. POLITICAL CARICATURE
Cologne Hänneschen Theatre. May 1848. Sketch by Kleinenbroich

In Hamburg the hand-puppet theatre flourished in the middle of the nineteenth century, when Johann E. Rabe was yet a boy. It owed much to the actor Küper, who was specially skilful in giving his puppets characteristic local touches. He abandoned the hunchback and the large nose which the Hamburg Kasperle had inherited from Punch, and gave him instead of those the face of a Hamburg workman with a corresponding dress—a red jerkin trimmed with yellow, blue trousers with yellow stripes, a white collar, and a blue peaked cap. Küper made his puppets, to which belonged Kasperle's partner, Snobelbeck, speak in the Hamburg dialect; indeed, he even set out to present all the five

FIG. 60. POLITICAL CARICATURE

Cologne Hänneschen Theatre. May 1848. Sketch by Kleinenbroich

The man with the broken sceptre and the bottle of liquor is supposed to be Frederick William IV.

different forms of speech in the town which corresponded to the various parishes. This attention paid to the Low German speech was of considerable service to Rabe. Küper died in 1893, and when on his death-bed he ordered his puppets to be burned.

Where the showmen lacked ability to give their shows the support of local characteristics their performances were confined to a repertoire which had to count on the lowest instincts to ensure a certain receipt. There was always Faust's descent into hell, together with many dramas based on pieces played on the regular stage—*Das Kätchen von Heilbronn* (*Kitty of Heilbronn*), *Der Freischütz*, *Alpenkönig und Menschenfeind* (*The King of the Alps and the Enemy of Man*), *Die Reise um die Erde in* 80 *Tagen* (*The Journey round the World in Eighty Days*), *Der Müller und sein Kind* (*The

FIG. 61. DOCTOR FAUST AS A VILLAGE BARBER AND MEPHISTO AS A CHIMNEY-SWEEP
Schmid's Marionette Theatre, Munich

Miller and his Child), *Der Fall Clemençeau* (*The Clemençeau Case*), and so on. Next came the blood-and-thunder element—*Die Totenglocke um Mitternacht* (*The Death Bell at Midnight*), *Die Leichenräuber von London* (*The London Body-snatchers*), *Der Mord im Weinkeller* (*The Murder in the Wine-cellar*), *Die Räuberschenke im Wiener Wald* (*The Robbers' Tavern in the Viennese Forest*), *Die Teufelsmühle am Wiener Berg* (*The Devils' Mill on the Viennese Hills*), etc. The announcement of scenic transformations proved always a special attraction. In 1899 a bill announced *Das Jochkreuz oder der Protzenbauer von Zehnerhof* (*The Cross on the Mountain Ridge, or the Insolent Peasant of Zehnerhof*), a rural folk piece, with the promise of "real water and rain," and in the same year *Der Lumpenball oder der verhängnisvolle Affe* (*The Beggars' Ball, or the Unhappy Monkey*) was advertised "with fireworks."

FIG. 62. SPANISH DANCERS
Schmid's Marionette Theatre, Munich

FIG. 63. KASPERLE AND HIS WIFE
Schmid's Marionette Theatre, Munich

Many of the old puppet-plays have been collected by Karl Engel, Kralik, and Winter. From these it is evident that the showmen, for example those of Lower Austria, while keeping to the old texts, fundamentally modernized them, bringing them

FIG. 64. VENETIAN MARIONETTE THEATRE, FIRST HALF OF THE EIGHTEENTH CENTURY
Victoria and Albert Museum

into line with the popular Viennese folk songs; they thus loved the songs of Girardi. The puppet theatre in the nineteenth century found only one genuine poetic writer—namely, Count Pocci, whose activities were determined by the fortunes of the Munich marionette theatre. In 1858 the Bavarian general Karl Wilhelm von Heydeck handed over a small marionette stage to Josef Schmid. The latter opened his theatre at the Maffeianger, and

FIG. 65. VENETIAN MARIONETTE THEATRE, FIRST HALF OF THE EIGHTEENTH CENTURY
Victoria and Albert Museum

in 1900 moved to a house which the city of Munich built for him in the Blumenstrasse. Schmid died in 1912, at the age of ninety-one. It is curious to notice that the manipulation of puppets seems to guarantee long life! This puppet stage owed a great deal of its success to the lovely compositions of Count Franz Pocci, who wrote for it in all forty-one pieces, which he collected in the six volumes of his amusing *Komödienbüchlein*. Pocci possessed an unconquerable humour which, although it is satirical, comes always from the heart, and not from the head. One cannot be annoyed with him, for the sarcasm which he loves to bestow on the official and scholarly world only tickles—

ITALIAN MARIONETTES
From Schmid's Marionette Theatre
Bayerisches National-Museum, Munich

MARIONETTES
Walter Trier

it does not wound. His Kasperl Larifari is a jester who draws his comic spirit essentially from the contrast afforded between his fully prosaic mind, based definitely on reality, and the world of fairy-tale in which he is set by the poet. Pocci brought the puppet-play to a high level; all that old innuendo is completely banished in his pieces. His dialogue and verse are as naïve and

FIG. 66. VENETIAN MARIONETTES FROM THE MUSEO CIVICO, VENICE
Eighteenth century
Photo C. Naya

natural as his way of thinking; he understands the people even while he remains above them.

Josef Schmid, who through a series of decades was much honoured by young and old as "Papa Schmid," had an extraordinary sense for the technical as well as for the emotional side of his art. The many thousand puppets which he made were charmingly constructed and clothed with great taste. For the parts spoken behind the scenes and for the working of the puppets he had a skilled company of seven men. Among his performances he included operas with choruses and solos. Rapidly the marionette theatre under his direction became for Munich a genuine cultural element such as may not be underestimated.

In Italy the marionette theatre for long has maintained a position not far different from that held by the regular stage. Its success was assisted by the theatrical propensities of the people. It is not therefore mere chance that the most successful

FIG. 67. VENETIAN MARIONETTES FROM THE MUSEO CIVICO, VENICE
Eighteenth century
Photo C. Naya

puppet-showmen in Germany, England, and France have come of Italian stock. By the middle of the eighteenth century Abbé Dubos saw grand opera produced in Italy by means of marionettes, and the custom of presenting large shows of this kind is still maintained. The *fantoccini* of Milan used to give performances of long plays, and are said to have produced something unique in the way of ballets. The *burattini* in Rome took into their repertoire those sentimental melodramas which were in great fashion in the nineteenth century, indulging too in ballets; for the latter the puppets were compelled to wear little blue tights, similar to those which the law enjoined on the living *ballerinas*. While the regular theatres in Rome were open only during the carnival, the puppet theatres were allowed to give shows the whole year round. There they took over Rossini's operas and presented, too, a number of realistic shows. Charles Dickens saw *The Tragedy at St Helena, or the Death of Napoleon,*

FIG. 68. VENETIAN MARIONETTES FROM THE MUSEO CIVICO, VENICE
Eighteenth century
Photo C. Naya

played by puppets, and was highly pleased with the performance. In the drawing-room the marionettes lost all their shyness. At a private gathering at Florence Stendhal witnessed a performance of Machiavelli's comedy *Mandragola*, an unequivocal avowal of libertinism, while in Naples, under the same conditions, he saw a political satire, at that time a truly dangerous undertaking.

The Italian Kasperle theatre developed in different centres characteristic types which personified the special nature of the inhabitants. In Milan there was Girolamo, in Turin Gianduja, in Rome Cassandrino, who did not hesitate to quiz heartily the almighty *monsignori* of the papal city. The Italians maintained their pre-eminence throughout the entire century; as late as 1893 the Prandi troupe won great applause in London. Of the Spanish marionettes it is recorded only that they remained true to their half-romantic, half-religious repertoire of olden times.

FIG. 69. VENETIAN MARIONETTES FROM THE MUSEO CIVICO, VENICE
Eighteenth century
Photo C. Naya

When, for example, *The Death of Seneca* was represented at Valencia in 1808 the blood flowed in streams (by means of red ribbons), while at the close the heathen philosopher went heavenward and made a Christian profession of faith.

The French puppet theatre could boast of pre-eminence over that of any other European country; eminent writers and artists espoused its cause or availed themselves of the opportunity afforded by it for realizing their ideals. The wandering marionettes played *Paul et Virginie* and *Atala* so long as the authors of these pieces were in fashion; then they turned to real events, such as *The Capture of the Malakoff in the Crimean War*. Apparently they did not rise above the level of the usual audience at the fairs. As in Germany and Italy the hand-puppet theatre created in diverse towns its characteristic types—types for which the *gamins*, with their colloquial tone, were taken as models. Thus,

FIG. 70. KASPERLE THEATRE IN ITALY
Engraving of the eighteenth century

Lafleur arose in Amiens—a figure said to have been invented by the workman Louis Bellette, who made him speak in the dialect of Picardy; from Lille came Jacques, and others sprang up in different centres. The famous Guignol at Lyons was carried on after the death of its founder by Louis Josserand and his family, but the most renowned French 'master' of the puppet-show in the nineteenth century was Anatole Cressigny in Paris. He was an artist in the full sense of the word. He himself wrote the *scenario* of his pieces, improvising the actual words during the performance; he carved the puppets' heads with his own hands; and he is said to have been such a brilliant player that he could speak dialogue in twenty different tones. He died in 1893. In France the puppet-show was a popular amusement. At

FIG. 71. KASPERLE THEATRE IN VENICE
Etching by Zompini. 1785

Paris in 1874 there were ten booths, or 'castellets,' measuring about 2 m. square, each equipped with twelve to fourteen puppets. These were stationed in the Champs-Élysées, the Luxembourg gardens, and the Buttes Chaumont, and entertained their public with cudgellings, generally meted out to all representatives of civic authority. Rabe states that a puppet-show in the gardens of the Tuileries before 1870 could count on Sunday takings of 400 francs; Anatole made as much as 100 francs daily.

A great, indeed a passionate, lover of the hand puppets was George Sand, who in one of her novels, *L'Homme de Neige*, gives the preference to them when compared with the stringed marionettes on the ground that the latter produce a less satis-

FIG. 72. KASPERLE THEATRE IN THE RECEPTION ROOM OF THE CONVENT (DETAIL)
J. Guardi
Museo Correr, Venice

factory impression in that they have a resemblance to human beings. The poetess established a complete puppet theatre in her *château* at Nohant in 1847: its history she has related in her *Dernières Pages*. Her son Maurice carved the heads, carefully but crudely, with close observation of everything that came within his experience. They were painted in oils without varnish and provided with real hair and beards; their eyes were of glass or merely indicated by black varnish, with a nail as the pupil—the latter is said to have been the more effective. By 1872 there had been presented at this theatre 120 plays, involving the use of 125 puppets, all clothed by George Sand. Generally only a *scenario* was prepared, the dialogue being improvised. The various items in the repertoire were published in 1890 by Maurice Sand; these included many parodies and skits on popular authors of the period. Another private puppet theatre like that at Nohant was established by the famous singer Duprez in 1864 at his estate in Valmondois. He made his puppets perform operatic travesties, and was given permission to produce them before the royal couple in the Tuileries. In 1861 Duranty opened in the gardens of the Tuileries a puppet theatre for which the sculptor Leboeuf made the figures. He wrote his own plays, which he published in a collected edition in 1880. They were, however, much too highbrow to be appreciated by the crowd. They had no success, and the little booth disappeared. Strange to relate, the Théâtre Érotique de la Rue de la Santé, established in 1862, fared no better, in spite of the fact that it was indecent enough to suit Parisian taste. Tissérand wrote the cynical plays presented there, and Lemercier de Neuville made the puppets. Although Henri Monnier, Théodore de Banville, Champfleury, Paul Féval, and Bizet supported this little theatre, it lasted for only one year. By 1863 it also had vanished. The publisher Poulet-Malassis wrote its history, for which F. Rops, whose talent well suited his subject, provided the illustrations. In the year that the Théâtre Érotique closed its hardly opened doors Lemercier de Neuville started a puppet theatre of his own—the Pupazzi. At the start his figures were merely flat, sharply silhouetted puppets, but afterward, with Gustave Doré's assistance, he replaced these by rounded and clothed hand puppets. Lemercier manipulated them—played, spoke, and sang, besides himself writing the plays. These were exceedingly witty comedies, characterized by a satirical tone, in which appeared various popular public characters of his day, such as Thiers, Jules Favre, Victor Hugo, Dumas Fils, Émile de Girardin. In all he composed 120 plays of this kind.

An entirely different type of theatre was projected by Henri Signoret. He made puppets worked by strings from below. The figures ran on deeply grooved rails, the heads, arms, and legs being set in motion by strings passed through the body of the puppets. Each figure had to be controlled by a mechanic, with another person singing or speaking. The scenery was painted by Rochegrosse. The inventor of this show had his first great success in June 1888 at the Petit Théâtre with *The Birds* of Aristophanes. His aims then flew high; he desired to perform more of the world's classics, making his puppets give performances of Cervantes, Molière, Shakespeare, and even Roswitha, but the public did not show much appreciation of his efforts. When the charm of novelty vanished Signoret was forced to close his theatre, in 1892. Two artists who called themselves Dickson and John Hewelt, but who in reality were two brothers of French extraction named Alfred and Charles de Saint-Genois, also invented special puppets for their productions. Alfred created a figure which was attached to his body, so that he could use both hands for manipulating it, and Charles made marionettes the strings of which were moved both from above and from below. The puppets danced like the Sisters Barrison or the

FIG. 73. THE MARIONETTES OF HENRI SIGNORET
Behind the stage. Paris. 1892

Otéros and spoke like Yvette Guilbert, and Maindron could find only one thing wrong with them, that they were too exact and left no opportunity for the unexpected. Both brothers had been stimulated by the English illusionist Thomas Holden, whose puppets were a combination of marionettes and automata. They behaved with such vitality as to create complete illusion, but in Lemercier's judgment their technical perfection was a fault, for they appealed to the eye and not to the soul, and thus possessed no individuality.

The Walloons are great friends of the marionettes. About 1900 there were fifteen puppet theatres in Brussels alone, some with comparatively large stages and often with hundreds of puppets. They presented countless sets; the costumes were rich, single puppets costing between thirty and forty francs. They played still the old legends, such as *The Four Sons of Aymon*, alongside plays by Maeterlinck. The puppet-figure here corresponding to Hanswurst is Woltje, a contraction meaning 'little Walloon,' who is introduced into all the plays, be they comic or tragic.

VIII

THE SHADOW THEATRE IN THE ORIENT

HITHERTO we have been tracing the development of the puppet stage in its various aspects up to the threshold of the modern period. Before going further and attempting to give an account of its present position we must cast a glance at the non-European puppet theatres, which unquestionably have had a real influence on the marionette art of the Old World.

First of all comes the shadow theatre, which originated in the Far East, and entered Europe during the rococo period, at the time when China was the latest fashion. The shadow stage does not deal, like the marionette theatre, with rounded figures, but only with their shadows. Its technique is closer to the film than to the puppet theatre, since the art of the film also works, not with three-dimensional objects, but only with their two-dimensional representations. As a form of artistic expression it stands very high. "It is the art form," says Georg Jacob very prettily, "which approaches nearest the poet's dream, the creative power which reaches consummation in a waking dream; it can therefore reflect the poetic conception in all its freshness and original form, vainly striven after otherwise." Its origin is to be sought in China, where, during the Han period, in the reign of Emperor Wu (140–86 B.C.), it arose out of magical celebrations. Its scope embraces all possible incidents of the natural and supernatural worlds, intermingled with much grotesque humour and riotous fantasy. The figures are typical of the subject-matter. The good characters have human faces, the evil characters have devils' masks. These were not treated as pure silhouettes; the faces only were outlined in black. The ancient figures were made of bone or horn, transparent and painted; the modern, according to the description of Carl Hagemann, are made of stiff, oiled paper of a golden yellow colour; the bodies are cleverly built up out of a number of small planes, painted in transparent colours and superimposed on one another. Their colour effect, according to the same authority, has a subtle delicacy; for they are said to glow like stained-glass windows. These puppets are about 30 cm. high; the arms are movable at the shoulders, elbows, and wrists, the legs at the thighs and knees, supported and operated by means of bamboos strengthened by wire. The representation of scenery is not without charm: hills, rocks, trees, houses, and pagodas—all are introduced. The ancient

FIG. 74. TRANSPARENT COLOURED CHINESE SHADOW-FIGURE: WA-HI, THE PRIEST OF THE TEMPLE ON THE GOLDEN MOUNTAIN
Collection of Carl Niessen, Cologne

texts, of which Berthold Laufer, Wilhelm Grube, and Krebs have published sixty-eight different versions, are of importance for the study of the Chinese language and people and for the history of their culture and literature. Laufer is of the opinion that the shadow-play represents the highest artistic level which dramatic representation ever reached in China, in reference, of course, to a period in the distant past. According to Carl Hagemann, it was once the refined toy of the cultured, an art for the learned, whereas to-day it is merely a hollow relic. People do not now know what to do with the figures; they cannot work them; they are stiff and inflexible or else aimlessly flop about with all their limbs; when several characters have to be introduced the player is helpless. The Chinese shadow theatre no

FIG. 75. LEATHER FIGURE OF A SIAMESE SHADOW-PLAY
Staatliches Museum für Völkerkunde, Munich

longer has its own repertoire; it simply takes over that of the regular stage. It has no public, and the educated classes pay no attention to it now. The Völkermuseum in Berlin possesses a complete series of artistically made ancient Chinese shadow-figures which includes fifty-one heads which could be interchanged as desired.

Japan, which is so nearly related to China, also has a shadow theatre. The figures employed there are of a simpler opaque kind, but are, like the others, grotesquely conventionalized.

In Siam the shadow-play is of quite a different sort. There only scenes from the *Ramayana* are shown, but not by means of single figures. On the contrary, the whole scene is drawn on a skin and the contours perforated. Eight to ten, even twenty or more, persons move the skin to and fro in front of a fire, so that its shadow is thrown on a white sheet hung at an angle. Two speakers explain the scene which is being shown, and the whole is accompanied by instrumental music. Although these performances are given only at high festivals, especially at the funerals of notables, a jester is not absent. The Siamese shadow-plays have been influenced by the Javanese, but their figures possess individuality.

The true home of the shadow-play is Java—at least, there it has attained its highest and finest level. Originally it must

FIG. 76. JAPANESE SHADOW-PLAY FIGURE: A BRAMARBAS
Collection of Carl Niessen, Cologne

have been a part of the ancient Malayo-Polynesian cult of the Javanese. The representation of the shadows of the old revered heroes and ancestors shows a clear religious colouring; only in later periods has this disappeared and the shadow-play developed into a mere amusement. As all the *termini technici* of the *Wajang* are originally Javanese it is evident that it must have been an art which arose in Java, taking over from the Chinese shadow-play perhaps only a certain amount of stimulus. The *Wajang* is classified according to seven diverse kinds, but of these only two, the *Wajang Beber* and the *Wajang Purwa* in both of its forms —the older *Purwa* and the later *Kerutjil*—fall within the category of the shadow-play. The *Wajang Beber* is a kind of primitive

Javanese Wajang Figures

Wajang Figure (Java)

film, consisting of large sheets, 2 m. long and 50 cm. wide, on which the scenes have been painted, rolled on a wooden bar. Seven such rolls of pictures are used in one performance, which usually lasts about one and a half hours without an interval. The theme is generally a *pandji* tale, the story of a prince who experiences mythical adventures. An invisible speaker recites

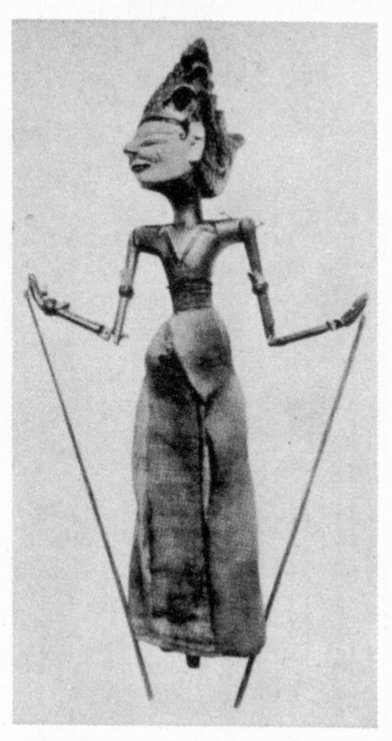

FIG. 77. WAJANG FIGURE (JAVA)

the story in a monotonous voice, without the accompaniment of music. At one time the *Wajang Beber* was a festival of high importance; later it was played in fulfilment of a vow, generally relating to the illness of a member of the family; and finally it sank into being simply a children's entertainment. When in 1904 Hazeu attended a performance at Jogjakarta he noted that there was hardly any connexion between the picture and the story, the reciter having forgotten the text because he so seldom was called upon for this task.

The *Wajang Purwa*, the native shadow theatre, has been known in Java since the first half of the eleventh century. It takes its themes from the *parwas* of the *Mahabharata*, *Ramayana*, and the Javanese cosmogony *Manik Maja*. The nature of the pieces is always romantic, generally with a religious flavour. In such cases they are supposed to serve the purpose of driving away or appeasing evil spirits, being then connected with magical and animistic rites. They introduce gods and goddesses, chieftains, princes and princesses, together with the jester Semar and his consort. Abductions are shown and battles with wild beasts, giants, wizards, and demons. The audience can never become satiated with these shows; a performance may last a whole night

FIG. 78. WAJANG FIGURES (JAVA)
(1) Semar. (2) Petruk

through, yet they do not weary. Sometimes, however, the show will continue for a whole week—pleasure and edification going hand in hand. The shadows fall on a flat white umbrella, and are cast by figures which have no equal in originality. As the Javanese people belong to the Mohammedan faith they are prohibited from making puppets in human form, and are thus limited to creatures of the fancy. The *Wajang Purwa* figures, therefore, resemble bizarre and fantastic ornaments in which a ghostly spirit is unintentionally introduced. They find an artistic parallel in certain reliefs from the *Ramayana* in the temple of Panataran in the Blitar district. The human body is in both transformed into an ornament; in the contours straight lines and edges are avoided; all is lost in curves. The figures are represented in sharp profile, with large noses and hair conventionalized in tail-like or spiral forms. Female hips are shown in front view, so that the costume can be thrust far out and the waist-line indicated as very thin. The hands and feet are shown in profile. The puppets possess hardly any likeness to humanity, but their expressions are always varied.

A centuries-old tradition has established certain definite types. Thus, a thin nose, flat brow, narrow, slanting eyes, and compressed lips indicate wisdom and high rank, while a short

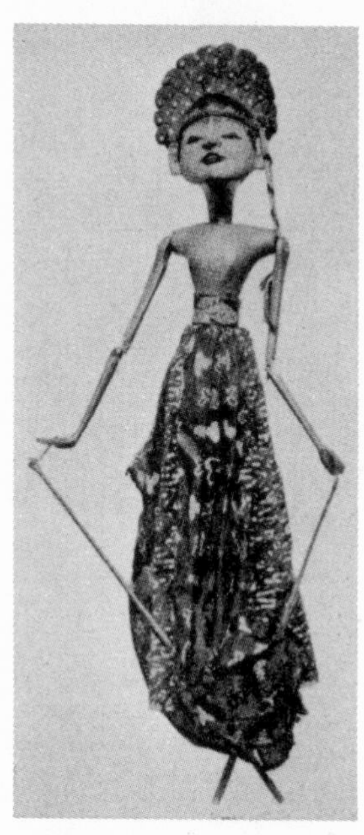

FIG. 79. WAJANG FIGURES (JAVA)
(1) Japeng Reges. (2) Prince

thick nose, a rounded brow, round eyes, and broad mouth characterize the hero of powerful strength. The arms can be moved at the shoulders and elbows, and each puppet is provided with an ornamental horn support which the manipulator can control with both hands. There are also figures with movable stomachs—there are even some with exaggerated movable genitals, the phallus often consisting of nine or ten parts, with the glans shaped like a bull's head. Although the style of the figures has been established by tradition, the variety within the narrowly marked boundaries appears unexpectedly great; some *Wajang* shows, indeed, include about two hundred puppets. Originally the plays were designed for men alone; only later were women permitted to witness them, and then only when they sat at the side of the showman, separated from the men. Thus, each of the two sexes sees the puppet from a different side, and this has led to the fact that the figures, originally intended only for the throwing of shadows, are painted and gilded on the side occupied

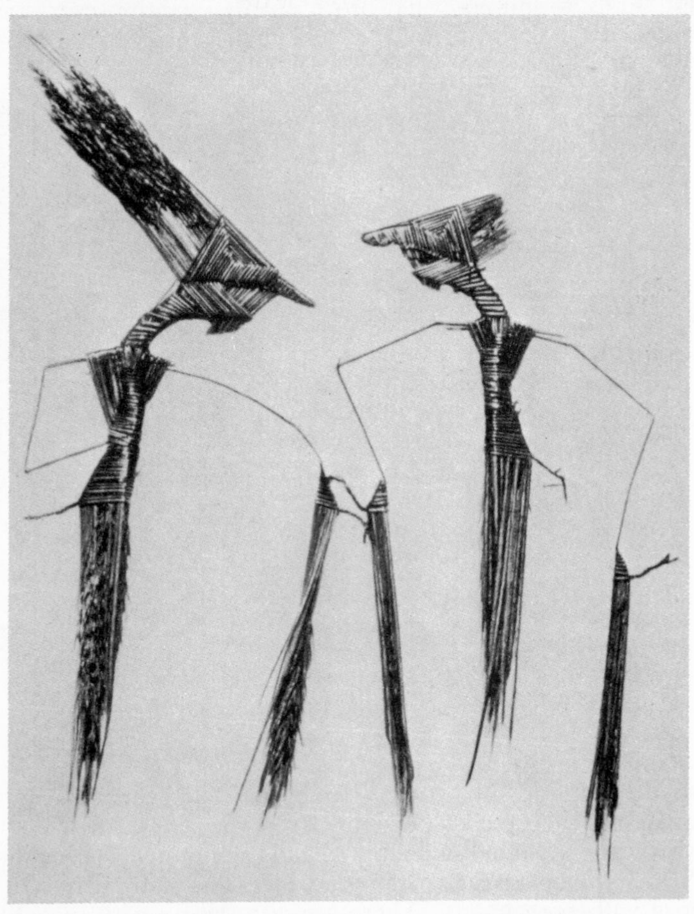

FIG. 80. WAJANG FIGURES MADE OF REEDS (JAVA)

by the showman. The upper part of the body, which is naked, as well as the arms and legs, is gilded; the hair and beard are painted black; the filigree work is coloured red, white, and dark blue. These *Wajang* figures are made of dried and smoothed buffalo hide, an art in which great skill is demanded. The tools consist of a small hammer of tamarind wood and about twenty-five to thirty different little chisels; the time taken over the work

FIG. 81. WAJANG FIGURES (JAVA)

FIG. 82. WAJANG FIGURES (JAVA)

is unlimited, several days being spent in preparing a single puppet. They are consequently very expensive; according to Gronemann, in 1913 one figure, as yet unmounted on supports and neither painted nor gilded, cost sixty Dutch guilders. This

FIG. 83. JAVANESE WAJANG FIGURES
See opposite
Collection of Carl Niessen, Cologne

is explained by the fact that they are chiselled with such inconceivable minuteness, often with an almost web-like effect, so that not only the form and facial expression, but even the ornaments on the head, neck, arms, and feet as well as the finger rings and the details of the dress are clearly defined. The showman, called the *dalang*, works them with oil-lamps over his head. He is player, speaker, and singer at the same time; he is the soul of the whole performance and must know everything. He has to learn by heart the endlessly long legends and be able to improvise when required; he has, too, to execute the traditionally conventional movements of the puppets. A bell orchestra, consisting of twenty to twenty-five men, accompanies the performance. The jester has his part to make the people laugh.

FIG. 84. JAVANESE WAJANG FIGURES
The shadows of the puppets
Collection of Carl Niessen, Cologne

Carl Hagemann gives the Javanese shadow-play a high place in the realm of applied art. He writes:

> The inconceivable refinement in the outlining and distribution of planes produces a great æsthetic pleasure when we witness these plays. The arrangement of light surfaces within the shadows reveals a delicate certainty in picturesque projection, and the moving lines of the arms create such a striking impression that one may remain in no doubt concerning the artistic worth of the *Wajang*. The most cultured Europeans themselves do not grow

weary of watching the whole night long with amazed wonder this dramatic black and white art distinguished by its originality and æsthetic power.

The Javanese *Wajang* spread not only to Siam, but also to Bali, Lombok, and throughout the Malay States to Sumatra and the mainland.

Among the Arabs the shadow-play was known from the

FIG. 85. WAJANG-WONG PLAYER (JAVA)

FIG. 86. THE MAN WITH THE PEACOCK

Islamic shadow-play figure from Egypt

eleventh century; in Persia it appeared at the beginning of the twelfth century. Omar Khayyám compares human life with a shadow-drama played in a box, the lighting of which is the sun, and in which we men come and go like puppets. In Egypt the shadow theatre flourished from the twelfth to the eighteenth century, with only one short interruption, when the orthodox Sultan Tschakmak (1438–53) ordered all the puppets belonging to this kind of play to be burned. His successors were of not so severe a disposition. Sultan Selim I, who incorporated Egypt in the Turkish realm in 1517, commanded the last Sultan of the Mamelukes to be hanged, and got this event celebrated by performances in the shadow theatres. The sole relics of dramatic poetry of medieval Arabia are three texts for shadow-plays which the Egyptian physician Ibn Danijal composed in 1267. During the Turkish domination the Arab-Egyptian shadow-play languished and was driven underground by the Turkish shadow-play, the players departing from Cairo for the smaller villages of the Nile delta. The texts, dealing with folk-lore material, were learned by heart by the players and handed down by oral tradition. In this way gradually traditionalized scenes were

FIG. 87. DAHABIYA ON THE NILE
Islamic shadow-play figure from Egypt

established with certain national types. About 1870 Hassan el Quasses founded in Cairo a new Egyptian shadow theatre for which he introduced new figures of the Syrian kind. The old figures had been made by cutting out pieces of leather, the impression being secured not only by the outlines, but also by the thin, transparent, multicoloured skin which was sewn round the bare flat figures. The making of these was a great art. Paul Kahle has unearthed a whole collection in Menzabeh, the latest specimens of which are about two hundred years old. As works of art these are to be compared with Gothic stained-glass windows, but the later ones, no longer indigenously Egyptian, show the whole figure as transparent. The figures used to-day (for the sake of economy often made only of paste-board) are ½ to 1½ m. high and have movable limbs. They are pressed, by means of a palm-leaf rod 1 m. long, against a tightly stretched white sheet attached to the side wings. The repertoire consists of lengthy pieces such as the *Alam u-Ta' adir* spoken of by Kahle, which includes material sufficient to enable it to be played continuously for the twenty-eight evenings of Ramadan. The crocodile play is that most popular in the Egyptian repertoire.

One feature is common to the Oriental plays, to the medieval mysteries, and to the melodramas of the eighteenth century: the comic figure, originally introduced incidentally to provide

FIG. 88. TURKISH SHADOW-PLAY FIGURES ("KARAGÖZ")

laughter for the audience, becomes in the end the pivot of the whole piece. A very good example of this is provided by the Turkish shadow-play. In the thirteenth century a Turkish word signifying 'shadow-play' made its appearance; in the sixteenth century this form of drama was popular in Constantinople, the sultans having their own private troupes, some of them highly renowned. The *Karagöz* originated in the seventeenth century; it was referred to first by the traveller Thevenot, who journeyed in Turkey between 1652 and 1657, and even in Oriental records it is not mentioned at an earlier date. From that time on, however, the shadow-play and Karagöz have become wholly identified; the comic figure has entirely pushed aside the other subject-matter. *Karagöz* means 'black eye,' which practically signifies 'gipsy.' Tradition has it that he was invented by a dervish in Brussa for the purpose of opening the Sultan's eyes to the mismanagement of his ministers, but Luschan assumes that the figure derives from Persia. He is the comedian and bearer of the title *rôle* of the shadow-play named after him. His most striking characteristics are his priapism and a very large turban, which he loses when he gets a cudgelling. By nature he is a clown, a mixture of Hanswurst and a jester, always a match for educated folks because of his mother wit, in speech more obscene than witty. His partner is Hadschi Eiwad, who represents the cheerful spirits of the better Turks. He also is a clown of lax morals, but not so rough and low as Karagöz, affecting an educated speech, and delighting in the mingling of foreign Arabic and Persian words in his conversation.

FIG. 89. TURKISH SHADOW-PLAY FIGURE MADE OF CAMEL SKIN
Collection of Carl Niessen, Cologne

Both these *rôles* are stock types, with which are associated a few others who evoke popular laughter—the newly rich peasant, the Jewish merchant, the hypocritical and deceptive dervish, the Frenchman who murders Turkish, the Armenian, the Albanian, etc. The more ancient figures are very carefully cut from camels' skin, and are rich in design and in the planning of their joints. The outlines are sharp; the features have individuality; the colouring is varied, but executed with taste. These are often genuine works of art, whereas the modern figures are badly designed: all proportion is neglected, the painting is flaring with aniline colours, the material used is, as already noted, often only paste-board. The Turkish shadow-figures are very flexible, some of them having their limbs movable at all the joints. They are fixed on thin wooden sticks, by which they are controlled. When the showman, the *karagödschi*, has several puppets to manipulate his job becomes very difficult, for he often must keep his hands free in order to mete out blows, rattle castanets, etc. The *Karagöz* was at one time the chief entertainment at Ramadan, but the authorities were averse to the play, and Luschan noted a strong decline in its popularity during the eighties of the last century. The plays themselves fully reflected the physical characteristics of Karagöz; they were generally improvised from a *scenario*, and moved from the uncouth to the vulgar. Since the showman always made use of the popular speech the Government prohibited this, in an endeavour to diminish the audience. Special applause greeted the vocal tricks of the *karagödschi* when he stuttered, nasalized his words, and imitated foreign dialects.

Luschan knew of about fifty Karagöz plays, of which thirty were printed, although the fact that they were so printed did not prevent improvisation. Caustic invective and low wit directed at events and persons of the day constantly appeared here. The *Karagöz* was so dear to the Turkish heart that eventually even

FIG. 90. SARANTIDIS, VAKALO, AND YANIDI
Greek " Karagöz " shadow-play scene
L'École Medgyès, Paris

plays taken from the West and translated—no matter how serious they might be—had to be mixed up with comic business before they could gain success. Under Turkish rule the *Karagöz* became popular throughout the Balkans, and passed over to Algiers and Tunis. In Algiers the French suppressed it, suspecting it to be a means for the dissemination of political propaganda. The pitch of indecency was reached in Tunis in the *Karakusch* (*The Self-satisfying Swine*), where such scenes were represented on the stage as the violation of men and women. It is hardly believable that the *Karagöz*, because of its character, did not find its way to Paris long ago; it would have been excellently suited to French taste! And if it had got there, what trouble would have been taken to get it across the Rhine! It is undoubtedly the one thing lacking in the stage life of the German capital.

IX

OCCIDENTAL SHADOW THEATRES

EUROPE became acquainted with the shadow-play through the medium of Italy, while the French brought it into fashion as *ombres chinoises*. Georg Jacob, however, has demonstrated that it is to be met with at a much earlier date in Germany, and England too may make claims to priority. Ben Jonson ends his *Tale of a Tub* with a puppet-play in five scenes which are presented behind a transparent curtain in the manner of a shadow-play. A speaker standing in front with a magician's wand explains the action. In the West the figures are always real pure black shadows, no use being made of the Oriental application of coloured pieces. From the middle of the seventeenth century there is frequent mention of shadow-plays on the German stage—at Danzig in 1683, at Frankfort in 1692—often along with marionettes. The theatre manager Ferdinand Beck, who appeared at Frankfort in 1731, introduced between the acts of his melodramas artistic shadow-plays. In the middle of the eighteenth century a certain Chiarini gave performances of *ombres chinoises* at Hamburg; the figures were attached to strings, at the ends of which rings were tied, the rings going on the fingers of the performer, who manipulated them as if he were playing a piano. The amateur theatres too used to present shadow-plays. In 1781 Goethe got a shadow theatre built in Tiefurt, he himself and

FIG. 91. CHINESE SHADOW-PLAY: "LE PONT CASSÉ"
French woodcut. About 1830

Einsiedel preparing the *libretti* for the performances. The subjects dealt with in these earlier experiments are not now known, for the shadow-play could only get a footing in Germany after it had met with approval in Paris. This type of entertainment seems to have been known in Germany at an early date, but the first record we get of it is in the correspondence of Melchior Grimm on August 15, 1770. In 1775 a certain Ambroise opened

FIG. 92. CHINESE SHADOW-PLAY: PARISIAN TYPES
Beginning of the nineteenth century

a theatre of this kind, which in 1776 gave performances in London. In it was displayed a shipwreck in the midst of thunder and lightning, together with various transformation scenes, including a bridge broken into pieces, a scene which from that time on remained popular in the shadow theatre. From 1784 Dominique Séraphin, with his *ombres chinoises perfectionnées*, was a formidable rival of Ambroise. The former's puppets indicated physical features and dress by thin light strips and were much praised. "The puppets," writes Thiéry, in his *Pariser Führer*,

> represent human deportment very naturally. They dance on a tight rope and execute character dances with the greatest precision. Beasts of all kinds make their appearance here and move in their own special ways, and neither the strings nor the wires which hold and manipulate them can be seen.

FIG. 93. PEEPSHOW AFTER AN ITALIAN ENGRAVING
Beginning of the nineteenth century

FIG. 94. LATERNA MAGICA
French engraving. About 1800

Séraphin was much run after, and bore in mind the temper of his period, for from 1789 on he presented only antimonarchical pieces. He died in 1800, but his theatre closed finally only in 1870.

The Romantics loved the shadow theatre as they did the marionettes. Christian Brentano, Achim von Arnim, Justinus Kerner, Tieck, Uhland, and Mörike worked with it, while Count Pocci wrote some pretty pieces for it and designed shadow-figures. Even such an experienced theatre man as Kotzebue

FIG. 95. HOW THE FIGURES OF THE SHADOW THEATRE ARE MOVED BEHIND THE SCENES
French woodcut. About 1840

could not resist its charm. "His large and small shadow-plays," writes Countess Julie Egloffstein in 1817, "are unique in their kind. He has grasped everything that such things possess of the beautiful, and understands the art of producing great things with limited means." In 1827, indeed, there was a regular shadow theatre in Berlin. The great success of the shadow-play between 1760 and 1830 is closely connected with the fashion for silhouettes, which was then at its height. These were worn as pendants, hung on the walls, painted on furniture and crockery; and in the shadow-play they were welcomed in movable form, accompanied even by speech and song. When the new art of lithography pushed the silhouettes aside, and still more when the mechanically produced photograph completely banished for a time all artistic treatment of such things, the shadow theatre also disappeared. In France Eudel, father of the writer Paul

FIG. 96. THE TEA-PARTY
Movable figures from a shadow-play. About 1830
Theater-Museum, Munich

FIG. 97. PICK-A-BACK

FIG. 98. PUNCH WITH A MASK
Movable figure from a shadow-play
About 1830
Theater-Museum, Munich

Eudel, was an artist who still worked for the *ombres chinoises* in a skilful way, but he stood alone. Germany possessed highly gifted designers of silhouettes, such as Konewka, but the shadow-play was forgotten.

Its modern revival is due to French artists. In the Chat Noir, a cabaret started by Rodolphe Salis in 1881, Henri Rivière began to improvise shadow-plays in 1887. Caran d'Ache,

FIG. 99. FROM RIVIÈRE'S SHADOW-PLAY "LA MARCHE À L'ÉTOILE"

Willette, Lucien Métivet, followed him without being able to rival his efforts. Rivière's art provided a fantastic fairy-tale for the eye, deeply poetic in theme, of peculiar beauty in form, the whole a dream which vanished even as one strove to capture it. The artist made use of light and colour to steep his scenes in a mood made arresting through its strange magic. Before him there had been nothing similar to this, and since the Théâtre d'Art vanished, in 1897, nothing to equal it has put in an appearance. Here there were great successions of scenes, such as the *Sphinx*, where the conquerors of all ages passed before the Sphinx, *La Marche à l'Étoile*, where the poor and lonely, beggars, shepherds, and slaves, followed the star of Bethlehem, *Clairs de Lune*, *L'Enfant prodigue*, in which the art of illusion reached its highest and most perfect charm. Rivière added powerfully to the impression created by his work through the utilization of light to emphasize the separate pictures. This was cast through coloured glasses which had to be controlled by ten or twelve men. The Nile landscape floated in a bluish-green twilight, Golgotha flamed forth blood-red, the Sphinx faded into a cold, misty grey, the combination of colour-tones always striking the proper psychological note. Through skilfully handled cutting and diminution Rivière secured astonishing perspective effects by simple means. Nothing is left of all his work now, but even

FIG. 100. FROM L. TIECK'S "ROTKÄPPCHEN"
Schwabing shadow-play, with coloured transparencies by Dora Brandenburg-Polster

the careful postcard reproductions of his scenes remain yet real things of beauty.

Henri Rivière knew how to present the natural alliance of poetry and painting, to create a fairy-tale theatre which in its possibilities left the regular stage far behind; but his followers did not possess his talents. Hans Schliessmann, who was a native of Mainz, but became an Austrian subject because of his long residence in Vienna, where he had won fame as an illustrator for the comic papers, collaborated with Caran d'Ache in producing shadow-plays at the Vienna Exhibition of Music and the Theatre in 1892, but these had no more than a temporary success. Chronologically the next were the "Elf Scharfrichter" in Munich, who in 1900 introduced shadow-plays in their clever artists' cabaret. In November 1907 Baron Alexander von Bernus sought to revive the shadow-play on a broader basis. This attempt resulted from the æsthetic endeavours of the literary and artistic circle to which Schwabing contributed such a peculiar tone several years before the War. "The shadow theatre," wrote Willy Rath on that occasion, "is intended for those tired of realism; in the shadow is revealed an external simplicity, the truly perfect obstacle to realism." As Bernus himself says: "The shadow theatre reflects in its purest form the intangible world of the waking dream." The stage here was a screen of white linen, 1·15 m. by 90 cm.; the puppets were 35 cm. high,

FIG. 101. FROM L. TIECK'S "ROTKÄPPCHEN"
Schwabing shadow-play, with coloured transparencies by Dora Brandenburg-Polster

and were brought forward, unseen by the spectators, in strips. They had movable limbs, but the manipulators used restraint in giving them gesture. For illumination the petroleum lamp was preferred to electric light, since the former made the shadow soft and full, and it was possible to graduate the power of the illuminant. The figures were designed by Rolf von Hörschelmann, Dora Polster, Greta von Hörner, Emil Preetorius, and Doris Wimmer. The performances lasted from twenty minutes to an hour and a half, several speakers being engaged in the show. Bernus himself, Karl Wolfskehl, Will Vesper, Paula Rössler, and Adelheid von Sybel wrote the plays, but performances were given also of pieces from older literature, such as Goethe's *Pater Brey*, Mörike's *Letzter König von Orplid*; Justinus Kerner, Tieck, Pocci, and Hans Sach were also represented.

FIG. 102. SCHWABING SHADOW-PLAY
Munich 1907. Prologue to the Turkish shadow-play of Rolf von Hörschelmann

FIG. 103. FROM WOLFSKEHL'S "WOLF DIETRICH UND DIE RAUHE ELS"
Designed by Rolf von Hörschelmann

FIG. 104. SCENE FROM "DIE SCHILDBÜRGER"
Otto Link. Decoration by C. Tenner

FIG. 105. SCENE FROM "HEILIGE WEIHNACHT"
Otto Link

FIG. 106. OLD GERMANY
Lotte Reininger

FIG. 107. OLD HOLLAND
Lotte Reininger

FIG. 108. OLD FRANCE
Lotte Reininger

FIG. 109. OLD ITALY
Lotte Reininger

FIG. 110. OLD SPAIN
Lotte Reininger

FIG. 111. FROM THE SILHOUETTE FILM "PRINCE ACHMED"
Lotte Reininger

FIG. 112. FROM THE SILHOUETTE FILM "PRINCE ACHMED"
Lotte Reininger

FIG. 113. FROM THE SILHOUETTE FILM "PRINCE ACHMED"
Lotte Reininger

FIG. 114. FROM THE SILHOUETTE FILM "PRINCE ACHMED"
Lotte Reininger

FIG. 115. FROM THE SILHOUETTE FILM "PRINCE ACHMED"
Lotte Reininger

FIG. 116. FROM THE SILHOUETTE FILM "PRINCE ACHMED"
Lotte Reininger

A spinet supplied the music, and both figures and scenery were executed with great refinement; the performances were distinguished and tasteful; the imagination of the audience was aroused to a high degree; but its financial success was not such as to warrant the continuation of this theatre.

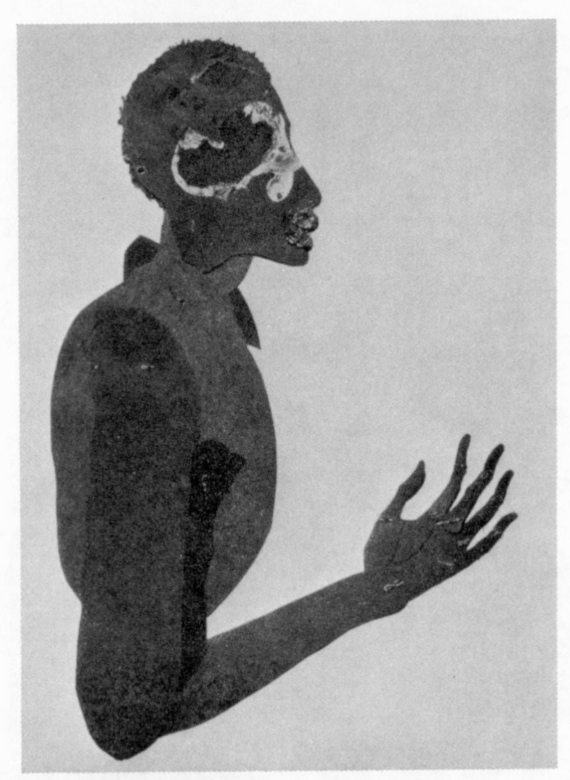

FIG. 117. SHADOW-PLAY FIGURE
Lotte Reininger

That such novelties should be copied in vulgarized forms in Berlin is comprehensible considering the ceaseless rush for sensations in that great city; such things commonly bear the mark of death on them at the very time of their birth. The Silesian shadow theatre which Friedrich Winckler-Tannenberg and Fritz Ernst opened in the Schiedmeyersaal at Breslau on November 15, 1913, looked as if it were to meet with a friendlier reception. They played *Doctor Faust*, a moral shadow-comedy in three acts adapted from the old *Faust*, Hofmannsthal's *Der Thor und der Tod*, Liliencron's *Die Musik kommt*, and other

FIG. 118. FAUST, SATAN, AND THE FOREST MAIDENS
From the shadow-play *Faust*. Design by Eugen Mirsky, Prague
By permission of the Deutscher Verlag für Jugend und Folk, Vienna

pieces, but the War brought this artistic and promising experiment to a premature end. Since then Bruno Zwiener has established a new shadow theatre in Breslau.

The peculiarly imaginative charm of the Javanese *Wajang Purwa* drew German artists too under its spell. Franz Bauer set up a shadow theatre in Bad Lausigk with figures inspired by the Javanese style; Bruno Karberg also gives performances at

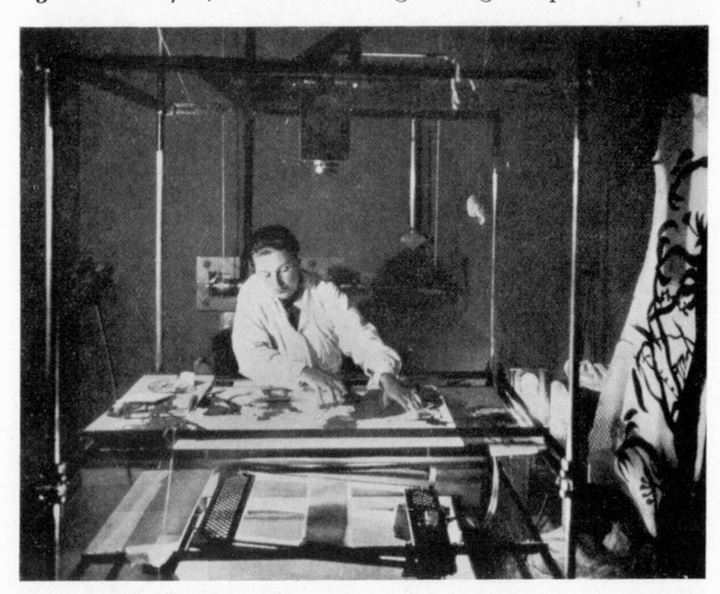

FIG. 119. LOTTE REININGER AT WORK

Hamburg with puppets which are supported, like the Javanese figures, on sticks worked from below. Käthe Baer-Freyer has been influenced by Javanese art in the making of her flat wooden figures, which she paints on one side and manipulates from below. The mystical charm conjured up by the shadow-plays has ever inspired the imaginative artists to further creative activities. Thus, in 1925 Hartlaub got Flaubert's *The Temptation of St Anthony* performed at the Kunsthalle at Mannheim, with shadow-pictures made by Wilfried Otto. Kurt Scheele in Frankfort produced shadow-pictures from the fables of Hans Sachs, songs by Richard Dehmel, and folk-plays; E. H. Bethge designs and writes for the shadow theatre; Eduard Maier goes on tour from Munich with his shadow-plays; Friedrich Winckler-Tannenberg in 1920 sought by means of his *Rakete* to introduce the *Morgenstern* shadow-plays to Berlin.

In the winter of 1925 Alfred Hahn at Munich made what seemed to be a most promising start with a shadow-play in colour dealing with the Nativity. Leo Weismantel has established an experimental shadow stage of peculiar interest in that the closely connected arts of the shadow-play and the film are here run together. In Munich Ludwig von Wiech had already put the shadow-play into a film, while Lotte Reininger with a silhouette-film showed what could be accomplished in this style. *Aladdin's Lamp*, the fairy-tale from *The Arabian Nights*, supplied the theme for *The Story of Prince Achmed* (*Die Geschichte vom Prinzen Achmed*). It was made by the Comenius-Filmgesellschaft, which between 1924 and 1926 must have taken 250,000 separate pictures, of which 100,000 were made use of. By the collaboration of the artist with Karl Koch, Walther Rüttmann, and Berthold Bartosch originated a work of art which on its production at the Volksbühne taught Berlin that this was no mere amusement for æsthetes, but represented new possibilities for the film. A short time ago the Prague silhouette artist Eugen Mirsky also tried this new path which the film has pointed out to the art of the shadow-play. In an exhibition held at Prague in June 1927 he showed silhouette figures which revealed in a peculiar mingling of the arts of the silhouette and the film a new phase, seemingly full of possibilities for the future, of the cinematographic art. His pictures are of no common sort; the step which he has taken from the merely ornamental silhouette to the naturalistic silhouette at all events may open up new prospects to the all too realistic film.

X

MARIONETTES IN THE FAR EAST

WITH this survey of the shadow-play we have come up to the present day, but now we must once more cast a glance at the marionettes of the Orient, if only because of their influence on the Western puppet stage. In China the theatre in which movable puppets were displayed must have been influenced by the Greek marionettes, the tradition being carried thither through Central Asia. Chinese records naturally say nothing of this. According to tradition Yen-Sze invented marionettes about the year 1000 B.C., presenting these before Emperor Mu of the Chou Dynasty. These puppets must have been very artfully contrived, for they were accused of shamelessly casting amorous glances at the ladies of the harem; for that reason the enraged monarch condemned them and their director to death. The angry potentate was with great difficulty persuaded that he had to deal here with lifeless things. Other records attribute the introduction of the puppet-play to a considerably later date, placing its invention in the year 262 B.C. At that period the city of Ping was besieged and was in danger of falling into the hands of the enemy. The prime minister then counselled Emperor Kao-tzu to get gorgeously dressed female puppet-figures led about the city walls so that by that means jealousy might be aroused in the enemy's camp, in which was the wife of the Emperor's opponent. His plan succeeded; the besieger was so much taken up with the faces of the lovely women that his wife became disturbed, and would not be appeased until her husband raised the siege of Ping and ordered his army to retreat.

There were three sorts of Chinese marionettes—those moved by strings, those manipulated by sticks from below, and the hand puppets. The first were of wood and were called *kui-lui*; the last were made of leather and named *pu-tai-hi*, which signifies 'sack-play.' These puppets played, sang, and danced love-stories, myths, legends, etc., the usual theme being that of an abducted princess rescued valiantly. The jester always took a part in these plays. There were also special pieces which were performed only before the imperial Court. The art of the puppet-showman must soon have reached a high level, for Sir Lytton Putney, the English Ambassador at Peking, was amazed

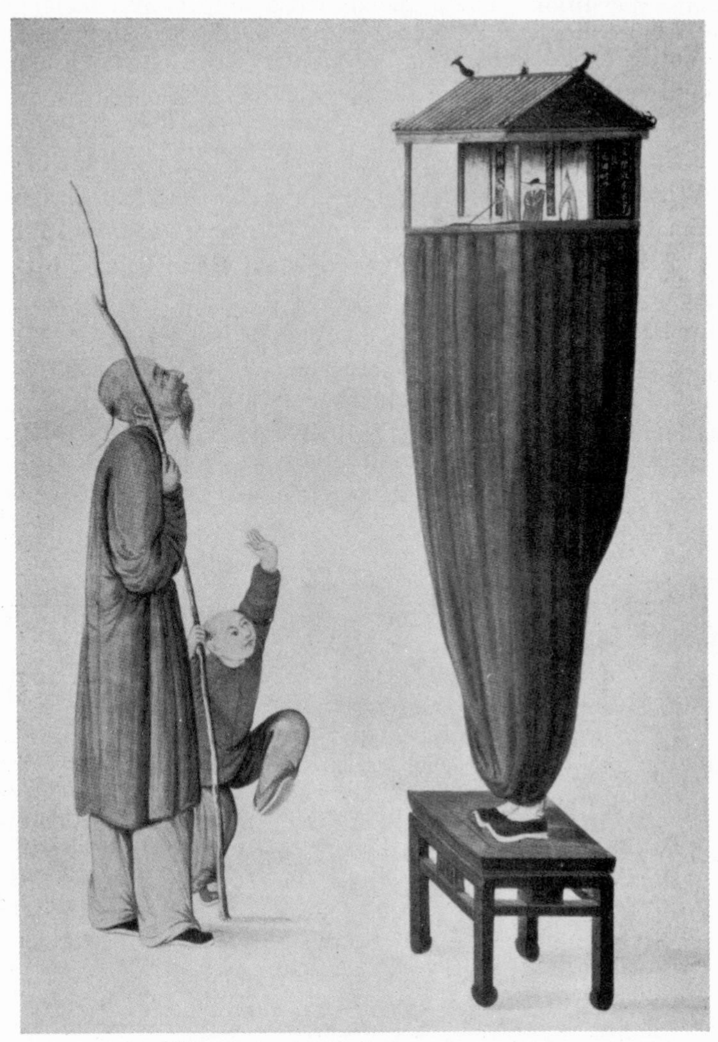

FIG. 120. CHINESE SKETCH, LATE EIGHTEENTH CENTURY
Puppet-play in the street
Victoria and Albert Museum

at the skill with which the showman manipulated his jointed figures and harmonized all their movements with the dialogue. In our own days interest in the puppets has apparently declined seriously. Hagemann, who was unable to witness a performance—so rare have the shows become in the larger towns—attributes this to superstition. The rounded plastic puppets were no longer popular, he thinks, since they were abused for magical purposes. Sprinkled with blood, they were supposed to turn at night into malevolent demons, on which account they were feared.

The puppet theatre in Japan occupies a position entirely of its own; there is none other like it in the world. Here men and puppets share the stage together and offer a complete substitute for the living actor. Carl Hagemann holds that:

> The Japanese puppets provide the most singular, the most genuine, and the most sincere theatre operating in the present-day world. What goes on here is theatre of the last and highest grade, the finest expression of art.

And a theatre director is certainly the person best qualified to judge of such things. The same authority continues:

> The puppets play better than real actors; they make a much better theatre than men. Their performances are more powerful artistically; because of the absence of curbing humanity they are presented symbolically with the highest intensity of expression; all reality has vanished.

These Japanese puppets are not marionettes of the ordinary kind, being worked neither by the hand nor by strings, but are rather mechanical works of art manipulated by men—the female puppets by three and their male partners by four individuals. One works the head, another the arms, and a third the rods. The puppets are dressed lavishly yet with refinement, the magnificent costumes being designed by such artists as Bunsaburo. In comparison with those who control them, dressed as these are in black or dark blue costumes and with their faces muffled up, the puppets, gay in colour and light-toned in features, create a powerful impression. They are about two-thirds or three-quarters life-size, those which take an important part in the play being larger than the others. Technically the male figures take precedence over the female, for they are given greater power of expression. The men can move their eyebrows, eyes, and upper lips while the women can only roll their eyes. The colour of the women's faces is snow-white, while that of the men's is flesh-tone; their features are sharply delineated and appear to Europeans close to the grotesque. Some of the heads are said to be very old, and cannot be reproduced nowadays since the secret of the lacquering process has been lost. Hagemann describes these heads as marvels steeped in inconceivable vitality.

Minister for Foreign Affairs — Minister for Home Affairs

Burmese Marionette Theatre

Staatliches Museum für Völkerkunde, Munich

THE EVIL SPIRIT NATT
Burmese Marionette Theatre
Staatliches Museum für Völkerkunde, Munich

They are set on ball-joints; fingers, wrists, elbows, and shoulders can be moved so that the figures can suitably manipulate their fans. Singers and reciters present the text, each figure having its own interpreter; only those of great renown for their art of delivery are permitted to deal with the dialogue between the puppets. The acting of the puppets is "inconceivably conscious,

FIG. 121. MOUNTED PRINCE
Burmese marionette theatre
Staatliches Museum für Völkerkunde, Munich

purposeful, inexorable." Paul Scheffer, who saw them in 1926, writes:

> The artists are so powerful in their manipulation and directing, in the profound godlike attitude they assume toward the creatures in their hands, that after a moment they succeed in making us forget their presence and are no more remembered. So at last, delicately and with a sure touch, they pour their life and whatsoever of thoughts they may have concerning life and human passions and human moods into the puppets which they hold. The black figures near them are their assistants. They all move delicately and always with dignity. Yet they do not hide, as we do, that they

have to make a thousand movements; the puppets are always openly surrounded by these mechanical devices, and through this they become uncanny. Before our eyes is clearly revealed the means by which the figure nods, rises angrily, throws a sharp glance to the side, or seems to laugh sneeringly; but, since all this seems to be part of its own life, it appears inevitable and indisputable. The boundaries between mechanism and inspired existence are, as it were, obliterated.

FIG. 122. HERMIT
Burmese marionette theatre
Staatliches Museum für Völkerkunde, Munich

This brilliant manipulation comes from the fact that the puppet-players are engaged in this trade of theirs from earliest youth; they dedicate their whole lives to the puppets. It is asserted that the living actors themselves have taken their histrionic style from the puppet theatres, and that from it they have borrowed their most impressive plays. This kind of theatre possesses a permanent stage in Osaka, founded about two hundred and fifty years ago by Takemoto Chikuyo Gidayu. He lived from 1651 to 1714, and was the most famous reciter of his time. In the middle of the eighteenth century it is said to have reached its culmination in poetical, musical, and technical form, but degenerated when the public developed a taste for the living actors. Alongside of this Ningyo-tsukai theatre is the puppet theatre called Ito-Ayatsuri, where genuine string marionettes give performances. This sort of puppet is, however, not much cared for by the public; apparently it was not indigenous to Japan, but an importation from abroad. In addition to these there are itinerant puppet-players in the form of wandering

FIG. 123. CLOWN (LEFT). MAGICIAN (RIGHT)
Burmese marionette theatre
Staatliches Museum für Völkerkunde, Munich

singers who hang their boxes round their necks and make their puppets dance on the top. The influence of the puppet stage on the regular theatre and *vice versa* has been traced by Carl Hagemann in Burma also. In the marionette theatres there the puppet first appeared as a reciter of prayers—a figure who appears at all public performances. The Burmese figures are made to dance just like the girls of the Pwe shows, but more grotesquely, wildly, and irregularly; the living girls, for their part, dance as though they are being pulled by strings. The puppets, with all their limbs movable at the joints, are supported on wires and strings, so skilfully that a good player not only can produce very agreeable and brilliant as well as naturalistic and grotesque movements, but can attain to great variety in the arrangement of positions. At the end of the last century Burmese marionettes gave a special performance at the Folies-Bergère, in Paris, which was received with great enthusiasm. A magnificent complete puppet theatre from Burma is in the possession of the Ethnographische-Museum in Munich. The wonderfully attired puppets hang on strings, the prince supported

FIG. 124. PRINCESS
Burmese marionette theatre
Staatliches Museum für Völkerkunde, Munich

by eleven of these, the princess by thirteen; the trunk of the elephant is composed of five movable pieces. The faces of the human persons are full of character; those of the evil spirits are caricatured. Realism and fantasy are thus blended in the Burmese plays. A prince and a princess are supposed to be pursued and fly into the jungle. There they are persecuted by demons, given counsel by hermits, and protected by good spirits—which bring all to a happy conclusion.

In Java, apart from the shadow-play, real marionettes are also to be found. One of the types is the *Wajang Kelitik* or *Kerutjil*, wherein appear, not the shadows of the puppets, but the puppets themselves. The plays have as their main theme the hero Damar Wulan and his deeds. Then there is the *Wajang*

FIGS. 125, 126. PRINCIPAL FIGURES OF A BURMESE MARIONETTE THEATRE
Staatliches Museum für Völkerkunde, Munich

Golek, which is played with rounded, wooden, dressed puppets, with loose and movable heads. Here for the most part are introduced heroes from the Mohammedan *Amir-Hambjah* cycle. This kind is said to have completely supplanted the others and to be much beloved by the spectators.

In India the puppet theatre as a popular amusement preserved its existence up to recent times. Now, indeed, when "Europe's whitewashed mediocrity has dominated the whole world," as Carl Hagemann says, "in a few years we shall not be able any more to find in India the genuine and the native. Like a grey rain-cloud, Europe's sobriety covers the ancient magnificence of colour. Soon all days will grow like eventide." The American film, with its dreary tastelessness, will be the death of native talent in India too. The pieces played on the Indian puppet stage were amazingly long; plays of seven to ten acts were no rarities. The stock piece, *Samavakara*, had, it is true, only three acts, but of these the first act alone lasted nine and a half hours, the second two and a half, and the third an hour and a half. As an afterpiece to this endless show a farce was presented in which the Indian Hanswurst, called Vidusaka, was given the chief *rôle*. In very early times in India—Pischel says in the eighth century—the attempt was made to banish this obscene jester from the stage, but without success.

XI

THE PUPPETS OF TO-DAY

We have traced the history of the puppet theatre up to the last century, a time in which its very existence was in question. It seemed then to be confined to the fairs of remote districts, and, although still retained in the nurseries, seemed rather to be tolerated there than encouraged. And just as it appeared to have died away completely, it showed that it still had life in it and still offered possibilities for artistic activity, on account of which it was seized upon from all sides. Many things contributed toward its revival. The primal impulse came from the professional artists. It will be remembered how at that period many painters, somewhat tired of the so-called 'fine' art, turned to 'applied' art, in order to find in creative work of this kind a source of satisfaction which the constant change of tendencies and the conflict of ideas denied them in the other. They discovered, accidentally, among the furniture, implements, and toys of the period of 'honest workmanship,' with which they sought to identify themselves, the old marionette theatre. This discovery seemed to them the happier in that the puppet-play apparently possessed that charm after which they all strove so passionately—the *naïveté*, the true simplicity of the folk. Here they were joined by those poets who had turned their backs on the prevailing dreary realism of the time in an effort to capture the romantic point of view. These were the days when there was a distinct turning away from Zola and a movement toward Maeterlinck. The dramatists greeted the puppet stage because it excelled the regular theatre in pure simplicity. Not the living actors, they opined, only the marionettes, were capable of expressing poetry without a distracting wilfulness; the human stage prohibits this, the puppets never. The marionette is naught but the expression of the artist's idea; the actor is always a man, and only too often his personality seems to place an obstacle in the way of true expression of a thought. Support from former times was found to strengthen this idea. Jean Paul emphatically demanded marionettes instead of actors for the performance of comedy. As early as 1839 Leman had asserted that a well-equipped marionette theatre would be the genuine native German folk theatre, and his contemporary Justinus Kerner had spoken out against living actors on the stage. The latter writes:

> It is peculiar, but to me at least the marionettes seem much less restrained and much more natural than human actors. The marionettes have no life apart from that of the theatre. With the marionettes and the shadow-plays the illusion is rather as if events

FIG. 127. PLUTO, PRINCE OF HELL
Last work of the puppet-showman Xavier August Schichtl, Munich (1849–1925)
Theater-Museum, Munich. Photo Hatzold, Magdeburg

> really taking place in one part of the world were seen mirrored here in small as in a *camera obscura.*

Not only did the writers of the older generation, such as Georg Ebers, Felix Dahn, and Adolf Wilbrandt, prove themselves enthusiastic friends of the puppet theatre, but they were joined by the younger men—Gustav Falke, Richard Dehmel, and others; such a successful dramatic poet as Ludwig Fulda could confess that "puppet theatres were once my highest ideal." Schnitzler, too, and Hugo von Hofmannsthal found the regular actors who interpreted their pieces too coarse and obstinate, and they cast longing glances at the puppet theatre. Even such a great actress (and director) as Eleonora Duse wrote to Vittorio Podrecca: "I envy you. I too should have liked to be the director of a puppet-troupe. Your actors do not talk, but obey; mine talk and do not obey." The most pertinent thing ever said concerning the connexion between the marionettes and the actors we owe to Bernard Shaw. The words which he wrote to the Italian puppet-showman Podrecca are reproduced in the introductory note at the beginning of this book.

FIG. 128. KASPERLE
Hand puppet. Carlo Böcklin

To the artists and the poets came the æsthetes, with their "homesickness for childhood," as Carl Niessen has put it so admirably. This longing for a distant paradise led men back, as in the eighteenth century, to the Far East, where they met with the puppets and the shadow-plays. To those desirous of reaching backward from the stylelessness of naturalism to style and to powerful form the marionettes must truly have been welcome. Instead of representing the human body and its movements naturalistically, they deliberately diverged from the real proportions of physical nature. The true puppet is not an individual, but a type. Its centre of gravity lies in its lead-filled feet, corresponding neither with the natural nor with the optically apparent. This determines the nature of its movement, and it is precisely the stiffness inseparable from the marionette which gives it its character. The facial expression as well as the bodily movements is typically fixed; instead of being expressive of changing moods, it provides a kind of epitome of the part. The theme and form of the marionette-play have to harmonize with these features, and from this results the last necessity of the marionette stage. The puppets are passionless; they can only

FIG. 129. THE LITTLE ROBBER
Hand puppet. Carlo Böcklin

act; such pieces alone can be performed as are developed simply, without any kind of psychological content. Psychologically the theme of the play must be limited to the representation of types, the conspicuous qualities of which must be clearly indicated. The marionettes will not and cannot give life by themselves such as it is the aim of the living actor to present, but provide an artistic reproduction of that. The special rules on which the marionette stage is based, the possibilities of expression confined within narrowly marked boundaries, demand of the audience a specific exercise of its imagination and a strong faculty of illusion not granted to every one. Even the pedagogues joined in this chorus of manifold acclamations by which the revived puppet-plays were welcomed. Writing in 1904, Paul Hildebrandt observes:

The puppet-play demands by its nature the greatest participation on the part of the child, and hence is one of the best and most instructive of theatrical forms, that on which adults ought to bestow the greatest attention in the interests of the artistic education of our children.

Konrad Lange too was convinced that Kasperle and marionette theatres aided the child toward an appreciation of plastic

FIG. 130. THE DEVIL
Hand puppet. Carlo Böcklin

art and of drama, providing the boy with a bridge leading to the enjoyment of other dramatic art. This philosopher complains that in so many families no attention is paid to this subject, and considers this a serious neglect in the education of children. No other occupation, in his opinion, will lead children so surely to the very core of all artistic expression as the theatre. In every family Lange would wish to see in use a small puppet theatre.

Among the patrons of the puppet-play was also numbered the Kunstwart, an organization which supported all endeavours seriously made to raise German culture. Göhler, who acted as spokesman, would recognize in the Kasperle theatre the simplest and most fitting form of art conceivable for the childish understanding, strongly shaping and limiting all the extravagant dreams of a still unyielding fantasy. If domestic episodes are chosen for the subject-matter, there then arises a fine and entertaining opportunity for the appreciation of daily life and of familiar surroundings in a simplified artistic form. Then Kasperle, who truly knows the good and bad conduct of children, may with his exaggerations and his good humour transform into laughter some things which otherwise would turn to tears. Thus the Kunstwart pleads for the revival and greater cultivation of Kasperle in the nursery.

FIG. 131. OLD WOMAN
Hand puppet. Carlo Böcklin

Frankly, for this, besides goodwill and some money, talent would be required. Nothing could come of the way in which the Teachers' Union at the Leipzig Michaelmas Fair in 1912 interested themselves in the Kasperle theatre. "Kasperle as a medium of education" makes much less impression than the hero of the fairs, with all his coarseness and extravagant cudgellings; he is simply a bore. Kasperle was a grossly abandoned creature, no one can deny it—but all his pretensions against authority go back in origin to the primitive idea which brings to a good end the struggle of an intrepid man against inimical powers such as death and the devil. He conquers them since he possesses humour, in spite of it all remaining an incorrigible old sinner, for he is always craftier than his opponent. It is a

FIG. 132. FROM THE OPERETTA "DER TAPFERE KASSIAN," BY O. STRAUSS AND A. SCHNITZLER
Figures by I. Taschner
Marionette Theatre of Munich Artists

FIG. 133. FROM "WASIF UND ARKIF"
Executed by Leo Pasetti
Marionette Theatre of Munich Artists

falsifying of the character of this strongly sketched personality to attempt, as the Teachers' Union does, to reduce it to a mere speaking tube which admonishes children to be good—to behave well at school, obey the teachers, study diligently, and bring home good marks. That is simply a case of getting Beelzebub to cast out the devil. The puppet-play, as the Kunstwart stated in May 1914, may become practical pedagogy, but the poetry need not thereby suffer. If the stage is turned into a *cathedra* it

FIG. 134. SCENE FROM THE THIRD ACT OF THE OLD GERMAN FAUST PLAY BY J. BRADLE AND P. NEU
Marionette Theatre of Munich Artists

will diminish in value; the child as such has a right to its imagination, and merely speaking about the child is not education at all. The puppet theatre should not only treat the child as a spectator, but should inspire it to participate in the play and to imitate; it should spur it on to active work of its own. If that were to succeed, then the puppet-play would be once more, as in olden days, one of the most popular forms of art—an art for which one can justly assert that it has roots in the folk, a thing which can be said of none of the fine arts at the present day.

One advantage of the puppet theatre is that it does not require the expensive and elaborate equipment which the regular stage demands, that it can produce wonderful things with the very smallest outlay. There is no lack of guidance; Leo Weismantel, Erich Scheuermann, Philipp Leibrecht, Peter Rich, Rohden, and others have given useful practical instruction concerning its use. At Easter 1925 a "Child and Theatre" exhibition, organized by the Saxon Teachers' Union in Leipzig, demonstrated all

From "Wasif und Akif oder die Frau mit den zwei Ehemännern"
by A. T. Wegner and Lola Landau
Figures and scenery by Professor Leo Pasetti
Marionette Theatre of Munich Artists. Paul Brann

Scene from Adam's Comic Opera "Die Nürnberger Puppe"
Executed by Professor Jos. Wackerle
Marionette Theatre of Munich Artists

the possibilities of its suitability for practical educational purposes. Stress was laid here on the hand puppets, the manipulation of which demands less time, skill, and patience than the marionettes. The puppet theatre is well advanced in the matter of practical instruction—indeed, such instruction has been actually

FIG. 135. AKIF THE ROBBER
Marionette Theatre of Munich Artists

established by law in Czechoslovakia. There is only one danger: that commercial interests take control of the affair and, by offering all possible 'improvements', debar children from developing and exercising their own initiative. Puppet theatres have been brought into commerce as "legally protected" novelties; the trade exhibits them equipped with cycloramas and electric lighting, and thus possessing all that a child's theatre can do without. In opposition to this the 1912 exhibition of the Munich Women Teachers' Union showed happily how children can be

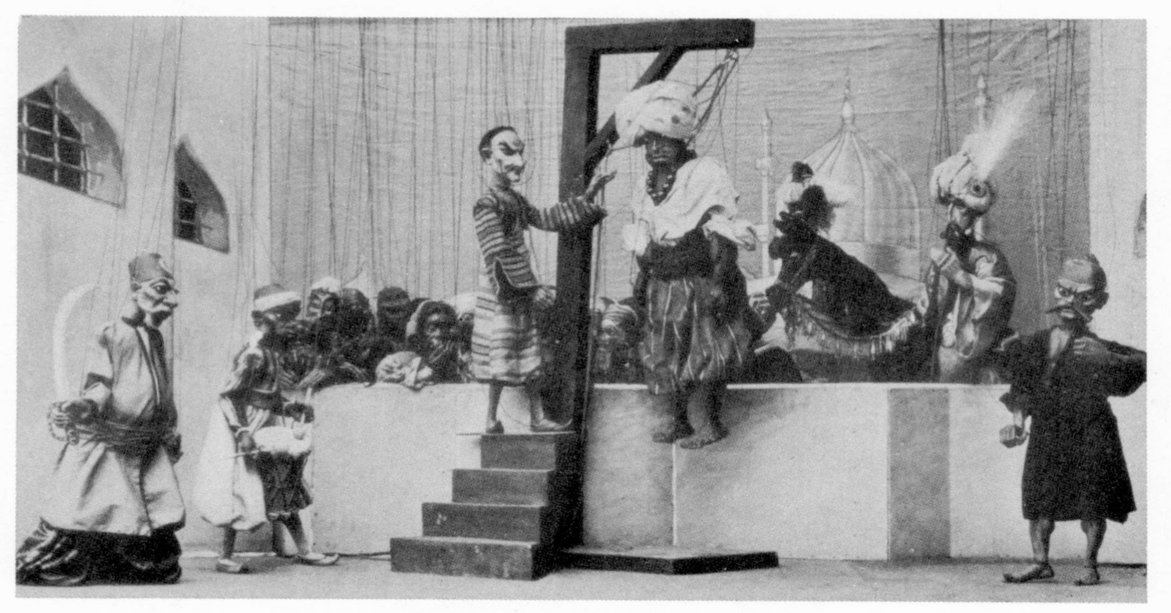

FIG. 136. GALLOWS SCENE FROM "WASIF UND AKIF"
Scenery by Leo Pasetti
Marionette Theatre of Munich Artists

FIG. 137. FROM "DER GROSSE UND DER KLEINE KLAUS"
Marionette Theatre of Munich Artists

inspired to work for themselves. Professor Bradl had carved most impressive heads from celery, turnips, potatoes, and radishes, and had dressed them in multicoloured pieces of sackcloth. These puppets gave performances against a simple blue background which could be copied by any child. Carlo Böcklin too, the son of the famous painter, possesses a peculiar talent for the puppet-show; he has made most original figures and

FIG. 138. SHEPHERD FROM THE CRIB-PLAY OF PAUL BRANN
Josef Wackerle
Marionette Theatre of Munich Artists

given improvised plays for his children. In his illustrated books on Kasperle he has indicated a new way in which the puppet-play can be introduced into nurseries. Beate Bonus, under Böcklin's direction, has collaborated in the writing and collecting of the texts.

The accountant Carl Iwowski, in north Berlin, owes to both the happy idea of his *Wandervogel-Arbeitsgemeinschaft*, by means of which he gives Kasperle theatre performances for poor children of this dreariest quarter of the German capital. He began by composing a Faust play and carving the figures for it himself. For this purpose he used peculiarly knotted roots which he painted, or else little blocks of wood which under his skilful

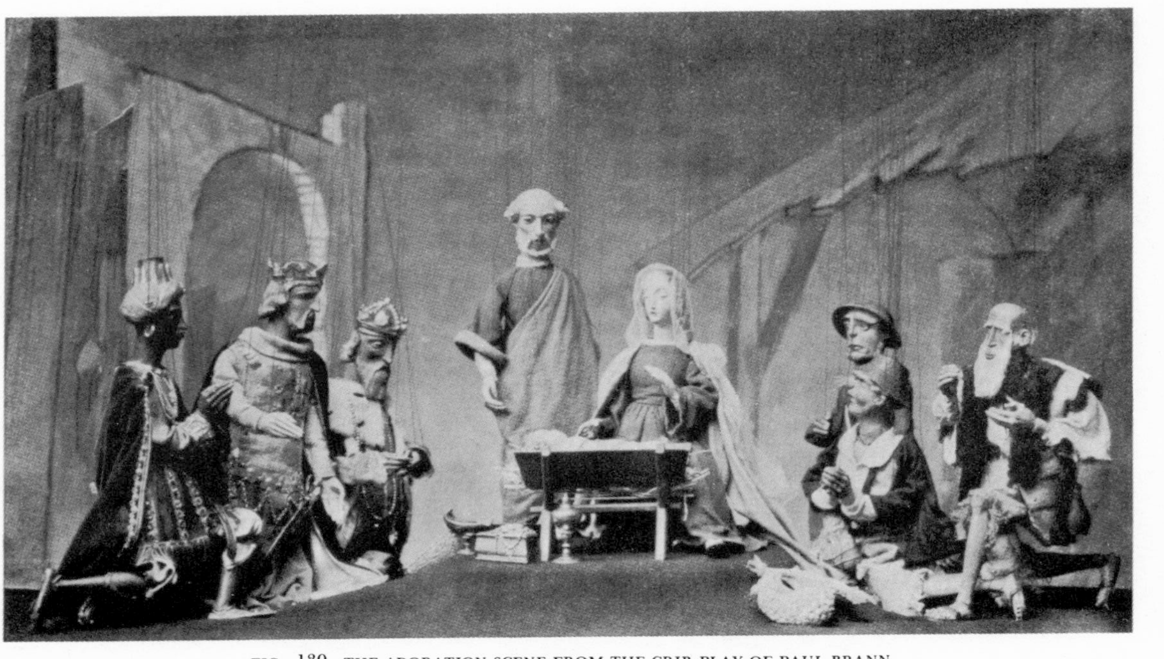

FIG. 139. THE ADORATION SCENE FROM THE CRIB-PLAY OF PAUL BRANN
Josef Wackerle
Marionette Theatre of Munich Artists

hands were transformed into wooden devils and other figures. He started in 1919, and since then has developed a completely new style in the presentation of hand puppets. He built a little theatre of his own design, freeing his puppets from the edge of

FIG. 140. THE POLICEMAN
Jakob Bradl
Marionette Theatre of Munich Artists

the stage, and so giving to his figures a physical materiality which hitherto they had been denied.

The renaissance of the marionettes is associated with two names, Paul Brann and Ivo Puhonny. The Marionette Theatre of Munich Artists, which has existed since long before the War, under the direction of Paul Brann, owes its world reputation to his puppets, the heads of which are executed by Ignatius Taschner, Jakob Bradl, and others, and to his excellent manipulation. On his miniature stage appears a harmonious unity of all arts, much more convincing than in the regular theatres; naïve dramatic work and pleasant settings are here triumphant, precisely because lifelikeness is not demanded from his stringed marionettes. Brann wishes, to secure success, that three intellects should collaborate in his performances—those of the

FIG. 141. THE STAR SINGER
Josef Wackerle
Marionette Theatre of Munich Artists

manipulator, the speaker, and the puppet itself. The puppet is thus considered an intellectual force by itself, and Brann believes that much is gained by giving it free play. His own performances, at any rate, fully justify this claim and permit him a wide-ranging repertoire. This includes the puppet-play of *Doctor Faust*, the peculiar *pièce de résistance* of all puppet stages, but he presents, in addition to that, plays by Mahlmann and Pocci, Ludwig Thoma, Schnitzler, and Maeterlinck; here, too, can be seen masterly performances of little comic operas by Pergolese, Mozart, Adam, Suppé, and Offenbach. At propitious moments the illusion becomes complete; we forget that these are puppets before us.

FIG. 142. HEROD AND THE DEVILS FROM THE CRIB-PLAY OF PAUL BRANN
Josef Wackerle
Marionette Theatre of Munich Artists

FIG. 143. THE EXAMINATION SCENE FROM "GOETHE"
Grotesque by Egon Fridell and Alfred Polgar. Figures by Olaf Gulbransson
Marionette Theatre of Munich Artists

Since the deaths of Taschner and Bradl, Josef Wackerle has become the special 'house' artist of this theatre. When in 1925

FIGS. 144, 145. SINGLE FIGURES FROM "GOETHE"
Olaf Gulbransson
Marionette Theatre of Munich Artists

Brann wrote a crib-play after the old style, fashioning it into six impressive "blessed Christmas" scenes, Wackerle made for him a setting, so real, so homogeneous, so spiritually refined both intellectually and physically, and with such perfect harmony, as could never be attained on the regular stage. With the help of this artist, and by his own excellent artistic skill in management,

FIG. 146. THE COMPETITION DANCERS
Totalia Meschuggerowska and her partner Fred
Ivo Puhonny's Artists' Marionette Theatre

the director produced a powerful "unforgettable impression" which the regular stage would be incapable of. Among folk types Wackerle likes to stick fast to the people of his Werdenfels native district, and on account of that he succeeds best in those peasant types such as were so excellently delineated in the figures of the jolly play of Big and Little Klaus. Occasionally other artists too have collaborated in the work of this theatre—Olaf Gulbransson, for instance, the well-known illustrator of *Simplizissimus*, who made the figures for a travesty of *Lohengrin*.

Ivo Puhonny has placed over his puppet theatre the motto: "A good marionette is of greater value than a living mediocrity"; and he has certainly created puppets which far surpass many actors. After long preliminary studies and travels in the Orient

FIG. 147. PUPPETS FROM A PIECE BY WEDEKIND
Ivo Puhonny's Artists' Marionette Theatre
Photo Delia

FIG. 148. "DIE BUSSE"
Japanese farce
Ivo Puhonny's Artists' Marionette Theatre

he established his little theatre at Baden-Baden in 1911; he also toured with his puppets, and in 1916 gave over the directorship to Ernst Ehlert, who paid visits to every part of Germany. The unity of expression in these puppets is due to the circumstance that the artist made all the puppets and the scenery with his own hands. He has executed several hundred character puppets, and has shown himself a master of the first rank in the art of

FIG. 149. CLOWN, WITH EXPRESSIVE MOVABLE FINGERS
Ivo Puhonny's Artists' Marionette Theatre

puppet-carving. This is by no means an easy task, for a good marionette must not be realistic; rather must it, if it is to create the proper impression, be conventionalized without becoming too rigid. Ehlert himself observes:

> It may be said that the marionettes stray into wrong paths when an attempt is made to make them as lifelike as possible. The puppet should not imitate the human actor; it has its own laws. It need not be beautiful; it has only to be characteristic.

On this basis Puhonny makes the heads somewhat too big for the bodies and exaggerates slightly the characteristic features, thus imparting to them a certain mimic quality which proves effective in the play of his stage lighting. The singular and

FIG. 150. THE PIANIST
Ivo Puhonny's Artists' Marionette Theatre

FIG. 151. PUPPETS
Ivo Puhonny's Artists' Marionette Theatre

personal charm of the Puhonny marionettes, wherever they have gone in Germany, has gained them warm and sincere friends; they threw the aged Hans Thoma into ecstasy. The scope of performance of these marionettes, according to Ernst Ehlert,

FIG. 152. FROM "PRINZESSIN UND WASSERMANN"
Richard Teschner

FIG. 153. FROM "KÜNSTLERLEGENDE"
Richard Teschner

is infinite; they are capable of dealing with any scenic requirement. They venture to attempt even tragedies and works not originally intended for the puppet stage, but perhaps they produce the greatest impression when they are engaged in their own particular sphere. Puhonny and Ehlert, too, cultivate the solo-marionette successfully. They base their work on types of the old itinerant marionette theatre which, thus improved, justify their right to existence as artistic creations. The Berlin Press described the "dancing Chinamen" as "the most perfect thing that has ever been shown on the puppet stage."

An artist who has long occupied himself with the problems of the marionette theatre is Richard Teschner in Vienna. Before 1909 he was settled in Prague, and there the idea of his puppet

FIG. 154. STARRY NIGHT
Richard Teschner

stage first came to him. The mysterious suggestiveness emanating from the marionette as from polychromic wax-plastic seized upon him, but the first naturalistically created puppets which he made, to serve as figures for Hofmannsthal's *Thor und Tod*, did not please him æsthetically; they were too close to nature in appearance and movement. Only after becoming acquainted with the Javanese *Wajang* did he attain, with its aid, to a style of his own. He retained its technique (the manipulation by means of thin sticks from below), but he changed the flat *Wajang* figures into rounded puppets in three dimensions, and instead of only projecting them on to a screen he put them on to an ordinary stage. Thus did he carry over the Eastern methods to the West, and he himself, besides making all the equipment, scenery, puppets, lighting, and music, wrote for his stage fairy-tales with a

FIG. 155. STAGE OF FIGURES, FROM "NAWANG WALAN," ACT I
Richard Teschner

FIG. 156. THE SLAYER OF THE DRAGON
Richard Teschner

genuine Hoffmannesque fantasy. These puppets, which Teschner made out of polished lime-wood and then clothed, are exceedingly refined and delicate, but under his direction they give "mimic dynamic performances of such a noble simplicity and quiet power that they far surpass human actors, and it may be deemed that they will be the originators of a new style of histrionic

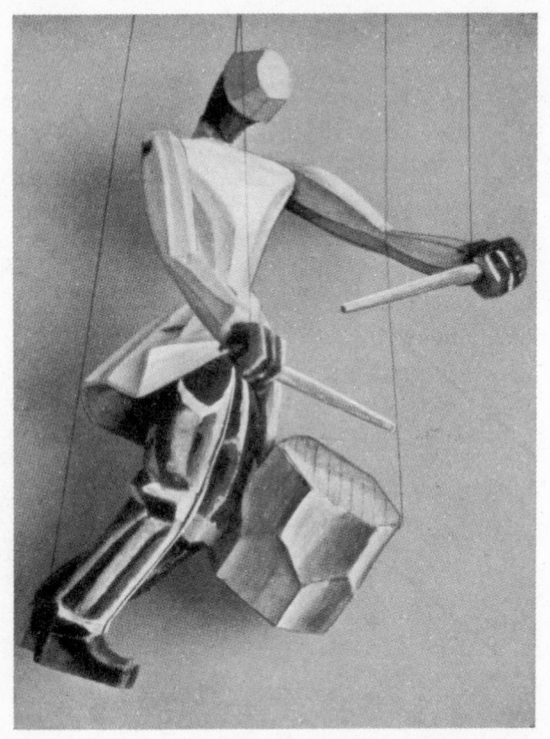

FIG. 157. MARIONETTE
Otto Morach. Solothurn

art." "Teschner's scene *Der Tod und das Mädchen* [*Death and the Maiden*] perhaps reveals the greatest possible refinement of the puppet-play," thinks Carl Niessen. The Viennese Kolo Moser was likewise attracted by the marionettes; in 1905 he made a puppet theatre with all the figures necessary for the performances given by Frau Lilli Wärndorfer in Vienna. At Zürich in 1918–19 a marionette theatre was established as an annex to the Kunstgewerbemuseum, in connexion with the First Swiss *Werkbund* Exhibition, and with the idea of testing on a small experimental stage the reforms for the regular stage proposed by Appia and

FIG. 158. SCENE FROM POCCI'S FAIRY-TALE "DAS EULENSCHLOSS"
Scenery by Max Tobler
Marionette theatre in Zürich

Gordon Craig. This marionette theatre works with flat and round puppets; it has found in Carl Fischer a talented carver and in Otto Morach a clever scenic designer. These artists collaborated in presenting a successful performance of *Faust*. The puppet-opera has been much encouraged in Zürich, for it was believed that by the aid of music greater possibilities of

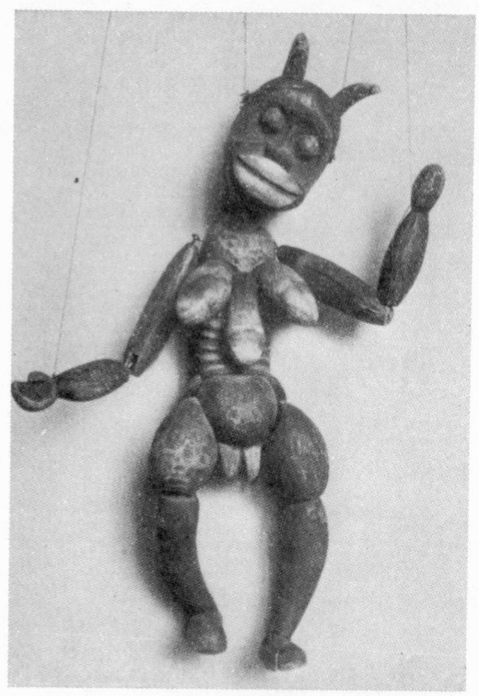

FIG. 159. DEVIL FROM "FAUST"
Carl Fischer
Marionette theatre in Zürich

illusion could be attained. "The variegated rhythm of music," writes Hans Jelmoli, "corrects the primitivity of the puppet, and changes the figure into an expressive exponent of every human passion." With the aid of these "animated marionettes" performances are given of Pergolese's *Livietta e Tracollo*, Mozart's *Bastien und Bastienne*, Donizetti's operas, and an opera composed for the puppets by Manuel de Falla—*Master Peter's Puppet-show*. The Zürich enterprise has gained success, too, outside its own town and canton.

How much living power, in spite of the popular craze for the

FIG. 160. SCENE FROM A PUPPET-PLAY BY TRAUGOTT VOGEL
Scenery by Ernst Gubler
Marionette theatre in Zürich

cinema, is still possessed by the puppet theatre, with all the poetic possibilities it offers, is shown by the Kasperle theatre

FIG. 161. SHEPHERDS FROM THE PUPPET-PLAY "DAS GOTTESKIND"
Carl Fischer
Marionette theatre in Zürich

founded by Dr Will Hermann at Aachen in 1919. He has created a puppet-play of a specifically Aachen character, and in Schängche (a diminutive of Jean) he has provided for it a counterpart to Hänneschen of Cologne, of such a characteristic and successful sort that this figure, animatedly reproducing the features of a canon, immediately became a freeman of his home

FIG. 162. DON QUIXOTE AND SANCHO PANZA
From *Master Peter's Puppet-show.* Otto Morach
Marionette theatre in Zürich

town. It is to the credit of the author that he has consciously turned aside from merely farcical elements, fully acquainted as he is, not only with the requirements of the puppet-play, but also with the life of his fellow-townsmen. But his merits were a disadvantage to him. Schängche could only express himself in the dialect of Aachen—an idiom of powerful expression within the walls of the old town, but outside of that incomprehensible. This has considerably limited his circle of admirers, and since naturally Schängche has not been without the usual lower-class competitors, counting not in vain on the baser instincts of the crowd, this enterprise, highly deserving of welcome, has not been

FIG. 163. VITTORIO PODRECCA BEHIND THE SCENES OF HIS TEATRO DEI PICCOLI, ROME

financially remunerative, and struggles with economic difficulties which unfortunately may perhaps force it to close its doors. To-day the things that find success among the German public have to come from America. There is indeed naught else that can appeal to them.

The most brilliant marionette theatre of the present day, introducing puppets of first-rate quality, is the Teatro dei Piccoli of Vittorio Podrecca, in Rome. This is completely Italian both in manner and in style, but it has given many performances in Germany, meeting with universal applause and providing lasting stimulus to other efforts. "The Teatro dei Piccoli is welcome," writes Pietro Mascagni. "Here is an artistic show of the first rank, which deserves the applause of all who admire artistic forms which express beauty and sincerity." "Lying between dream and reality," writes the Duse with reference to this theatre, "the marionette can be perfect when

FIG. 164. MARIONETTE
Teatro dei Piccoli of Vittorio Podrecca, Rome

it is guided by a soul." Podrecca employs a puppet *ensemble* which has no rival in our days. His five hundred marionettes, with twenty-three people at their service, present a repertoire of operas and plays with which they have charmed the whole world. Between 1913 and 1924 Podrecca gave eight thousand performances at Rome, Milan, London, Madrid, Buenos Aires, New York, etc. He himself came to the puppet theatre from the regular grand opera stage, which had failed to satisfy him, and

FIG. 165. FIGURES FROM AN ITALIAN FAIRY-TALE
Teatro di Ciuffettino, Florence

it goes without saying that unity of artistic expression can be reached successfully only on the puppet stage. Podrecca's marionettes are works of art by means of which the stage fantasies of great composers and of great poets are brilliantly interpreted in a manner which could not be equalled by any singer or actor. These puppets, which are unusually large (somewhat over 1 m. high), have been modelled by artists such as Caramba, Grassi, Montedoro, Angoletta, and Pompei. The stage directing is masterly. Carl Niessen writes of it:

> The certainty with which the figures move on the stage, with which they turn, dance, romp, play at ball with one another, reach for objects, pull out handkerchiefs elegantly from their pockets and put them back again, is such that no comparison with others seems possible. Yet the performance does not aim wholly at absolute illusion. It is extremely charming when the puppet so

FIG. 166. TERESA AND HER BROTHER CATCHING BUTTERFLIES

FIG. 167. MISS BLONDINETTE WITH THE MAESTRO CAPELLACCI

Teatro di Ciuffettino, Florence

seriously takes the trouble to do what living people do and yet remains wholly a puppet, a great romantic toy.

Music adds to the impression made by these shows. Of old composers Pergolese, Mozart, Rossini, Donizetti, and Massenet are here represented; an attempt was made to present even Richard Wagner's *Die Feen* (*The Fairies*). The operas are all curtailed, only selections of scenes being performed, so as to

FIG. 168. "DIE ORAKELTROMMEL"
Adolph Glassgold. Head by Kathleen Cannell. 1926
L'École Medgyès, Paris

harmonize with the spirit and conditions of a marionette theatre. Of living composers Respighi, to take one example, has written for Podrecca's puppets his *Dornröschen*.

Alongside of the Roman Podrecca must be mentioned the Florentine puppet theatre, Ciuffettino, of Enrico Novelli, with its witty shows and clever puppets.

In the United States Tony Sarg is held to be the most talented director of puppets. The marionettes with which he has appeared in the larger cities there were made by Charles E. Searle. There they gave performances of *Rip van Winkle*, *Don Quixote*, and similar pieces.

FIG. 169. SCENE FROM AN OPERETTA
Teatro di Ciuffettino, Florence

In Paris the École Medgyès devotes attention to the puppet-play. The clever marionettes here take their droll features from the skilful hands of Adolph Glassgold, P. A. Birot, Kathleen Cannell, and others.

The marionettes have created a genuine school of their own. Just as in Japan the living actors took their style from the puppets, so attempts have been made by Germany to carry over

FIG. 170. "DIE ORAKELTROMMEL"
Adolph Glassgold
L'École Medgyès, Paris

the technique of the puppet-play to the regular stage. In Dresden before the War, as Carl Niessen notes, the old farce of *Maître Pathelin* was produced in the marionette style. At performances of *Flohs in Panzerhause* the actors at the beginning were represented hanging on strings; in fact, in Erwin Fischer's Berlin theatre, which apes the Bolshevik style, actors and marionettes are said to have played together:

> The revolutionary Russian theatre must bring its whole being ever more and more into line with the manner of the puppet-play. The tragic clown Vladimir Sokoloff has attempted new artistic possibilities in his puppet-play—a theatre of absolutely musical dynamics.

The honourable position which the puppet theatre, thanks

FIG. 171. MARIONETTES
L'École Medgyès, Paris

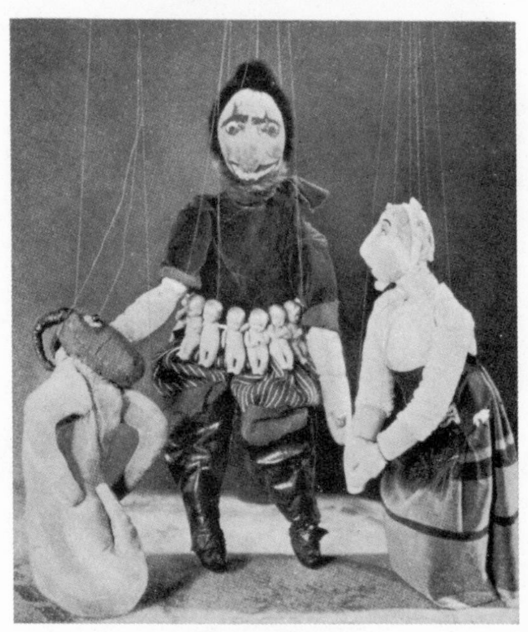

FIG. 172. LE PETIT POUCET
Pierre Albert Birot
L'École Medgyès, Paris

to the efforts of artists and connoisseurs, has held since the opening of the century is maintained sometimes with fluctuating fortune. Marionette theatres are founded, enjoy for a short

FIG. 173. KASPERLE THEATRE FIGURES
Max Pokorny and Tilly Gaissmaier

FIG. 174. KASPERLE THEATRE FIGURES
Max Pokorny and Tilly Gaissmaier

time a certain prosperity, and then disappear, generally from lack of means. They have not wanted either partisans or assiduity and goodwill, but whether it will be possible for them, with

FIG. 175. POLITICAL PUPPET THEATRE: PODBIELSKI, BÜLOW, BEBEL, MÖLLER
From the Jubilee Number "1000" of the *Lustigen Blätter*. Berlin, 1905

FIG. 176. MARIONETTES
Teatro Mazzini, Catania

their refined charms designed for a wholly intimate impression, to swim against the great tide of the film, with its tempting attractions to the public, is, alas! uncertain. A puppet theatre is

FIG. 177. MARIONETTES OF THE ARTS AND CRAFTS SCHOOL, HALLE
Photo Hatzold, Magdeburg

FIG. 178. R. WAGNER, A. MENZEL, BISMARCK
Kasperle theatre figures. Max Pokorny and Tilly Gaissmaier

easy to establish, but it is still easier, unhappily, to set up a cinema, and since the film is always based on a low taste, providing the greatest attraction to dull theatre-goers and flirting with the most evil instincts in the audiences, the mob will

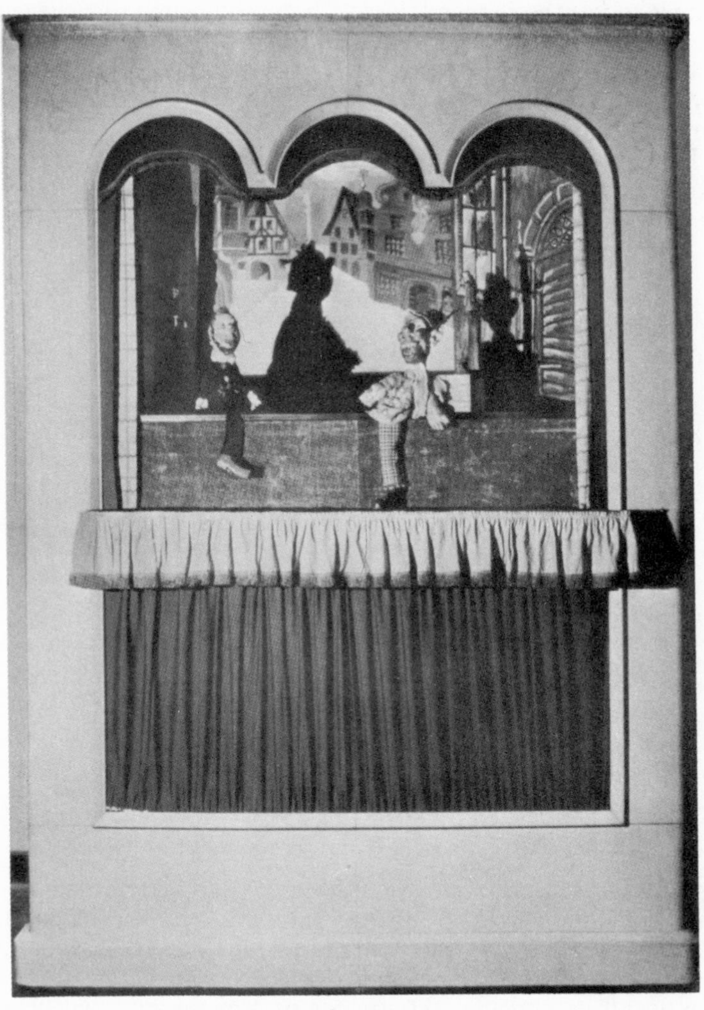

FIG. 179. MUNICH KASPERLE THEATRE
Max Pokorny and Tilly Gaissmaier

naturally give it the preference—of that there can be no doubt. Those who nowadays devote themselves to the marionette theatre must console themselves with the conviction that they are practising a refined and distinguished art, and thus serve the cause of good taste, which has always been a thing for the few alone, and in our present days is wholly an affair of the intellectuals.

Three factors which could be of high practical value have

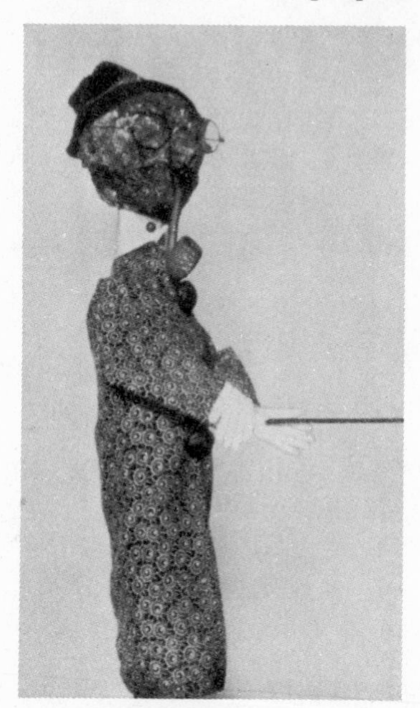

FIG. 180. FAIRY-TALE UNCLE
Puppet on sticks. Moulded in paper by Georg Zink

been recently placed in the service of the marionettes—the exhibition, the Press, and the university. At the German theatrical exhibition at Magdeburg in 1927 Dr Alfred Lehmann, of Leipzig, provided an unprecedented survey of the puppet-show in all its varieties, the greater part of the material being supplied from the well-known collection of Professor Dr Artur Kollmann in Leipzig. Since then there has not been even a small exhibition of toys which has not taken the puppet-show into consideration. Dr Alfred Lehmann also directed the *Puppentheater*, a periodical published by Joseph Bück in the interests of all puppet-showmen, which embraces within its sphere of interest both the history and the technique of the

PUPPET THEATRE
Walter Trier

puppets. The Union for Promoting German Theatre Culture has provided space in its official journal for a special puppet-theatre section. This paper, published by Lehmann and Schüppel in Leipzig, has been running since 1923. German high

FIG. 181. HEIDELBERG FLAT MARIONETTES
Above: Prospero and Miranda, from Shakespeare's *The Tempest*, and Austrian officer, 1750; below: fieldmouse and fox, from the fable-play by Georg Zink

schools, which for some years past have included the 'science' of the theatre in their curriculum, are also beginning to apply themselves to the puppet-play. The Institute of Theatrical Science at the University of Cologne established a puppet-play section in 1924, and no better person could have been chosen for its director than Dr Carl Niessen. He is theoretically and practically an accomplished expert in this art, and in his hands its scientific study is well provided for.

When, at the conclusion of our long wanderings through the puppet-shows of all times and of all races, we cast a glance at Germany of to-day we cannot find space to name all who deserve particular mention. A complete list it would be difficult to give, apart entirely from the fact that such a catalogue would have been in the end of very little use. An address book is not what is wanted here. Those interested in the puppets who are

FIG. 182. FIGURES MADE FROM ROOTS
Georg Zink

not referred to must not consider that the omission is due to ill-feeling or that it signifies any condemnation; they must realize that it is the result simply of lack of space.

It is but just to begin with Munich, which since the activities of the unforgettable Papa Schmid has remained the centre of serious artistic work in the sphere of the marionette theatre. The position it enjoys is due first to its artists, who have consistently devoted themselves more and more to this charming type of play, and secondly to the public, whose interest is strong enough to support several marionette theatres at one time. In Tölz the late apothecary Georg Pacher established an enchanting puppet theatre. In Graz the Kasperle theatre found many friends during the War. There Fritz Oberndorfer wrote clever texts which cast an amusing light on the commandeering of metal[1] objects, on summer-time, meatless days, etc., and satirized hoarders and War profiteers. Unfortunately this theatre has disappeared, along with the Viennese Gong. In Offenburg, Baden, the apothecary Löwenhaupt has sympathetically and lovingly devoted himself to the marionette-play. He belongs to the ranks of those who have most intimate knowledge of the puppet theatre, and he possesses a rich collection of objects relating to it.

FIG. 183. MARIONETTES
Marion Kaulitz. For Grimm's fairy-tale *De Fischer un syn Fru*

In the Heidelberg librarian Georg Zink Baden possesses another great connoisseur, collector, and promoter of the puppet-play. He works with his puppets both practically and theoretically in the interests of genuine folk education, supporting this most energetically by means of writings, exhibitions, and performances. His collections of objects relating to the history of the puppet-play and its literature, on which he has been engaged for many years, have been publicly exhibited in the theatre and music section of the town library, of which he is director. The marionette stage in Frankfort is famous for its colourful settings, for the clever manipulation of its puppets, and for its good, plain, characteristic dialogue. It indulges especially in the fairy-tale and in the solo-marionette introduced without words. In

[1] In reference to the War-time shortage of metal in Germany.

Brunswick Eduard Martens has devoted his attentions to the puppet theatre, and has brought his *Bodenkammerspiele* ('garret-plays') to a remarkable pitch of perfection. Many of his puppets were carved by Hans Pfitzner. The hand puppets in Hartenstein, in the Erzgebirge, owe their being to the merchant Max Jacob, who wrote the plays for them himself. He is reported to have succeeded not only in fairy-tales for children, but

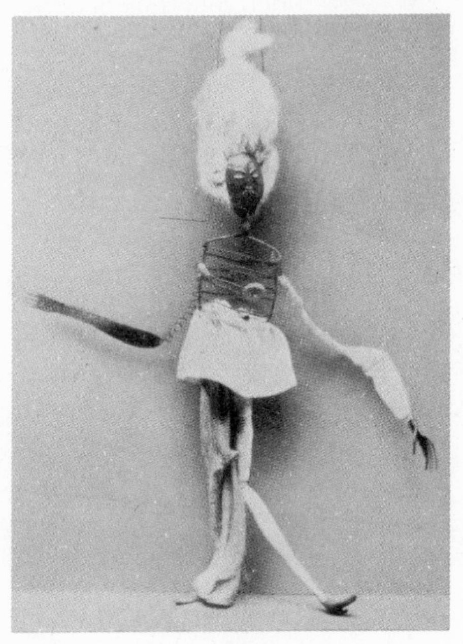

FIG. 184. MARIONETTE
Cläre Paech
Theater-Museum, Munich

also in satires of a topical sort. His puppets were made by Theo Eggingk, a young native of the Baltic Provinces, and were dressed by Elisabeth Grünewald. This little theatre and the young men who assisted in its work won considerable fame by their tours, in which they reached even to the fighting lines. They visited German Bohemia, East Prussia, Silesia, and Masurenland, endeavouring to bear to the Germans living there relics of the old precious folk art. Werner Perrey directs the Low German puppet-plays in Kiel, for which he himself writes fairy-plays, very beautiful and very charming, but heavily burdened with thought. As a poet Perrey demands too much—for the

FIGS. 185, 186. TRENCH PUPPETS
Made of limestone for a Kasperle theatre
Armee-Museum, Munich

FIG. 187. A TROUPE OF SOLO ENTERTAINERS
Josephine and Allardyce Nicoll

FIG. 188. SCENE FROM A MARIONETTE-PLAY: "AN INTERVIEW"
Josephine and Allardyce Nicoll

child's mind indeed he demands impossibilities. He makes his Kaspar speak too much and of too much. It is not possible to describe in one phrase the general tone of this theatre. In the Dessau Bauhaus experiments are being made with constructivist absolute marionettes, which consist only of abstract spatial forms created by means of wires, celluloid balls, rings, and beads. Wherever to-day creative artists are at work there the puppet theatre—be it with string puppets or only with hand puppets—finds sympathetic admirers to further its course. It is a pigmy among giants, but are not the noblest achievements in store for it?

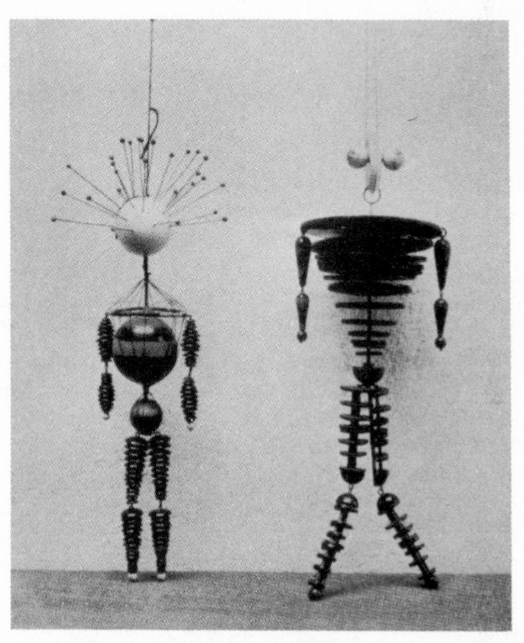

FIG. 189. MARIONETTES OF THE EXPERIMENTAL STAGE AT THE BAUHAUS, DESSAU
Theater-Museum, Munich

XII

HOW OUR MARIONETTE THEATRE STARTED

By HANS STADELMANN

YES, how did it really begin? Among the bombers it was, in the lines on the Dvina, that friend Barthels, who in civil life is a Munich actor, and I, a Munich painter, got into conversation about our childhood, and neither of us could say enough about the puppet theatres which each of us had possessed. And suddenly the thought was born: What a glorious thing it would be for our dull trench life if we could introduce such a theatre there! For us the work would not be too much if only we could succeed in winning the hearts of our Bavarian militiamen, and perhaps awaken in them some enthusiasm for this genuine folk art. Yet to build a marionette theatre in the midst of a military campaign, surrounded by a thousand hindrances—that task seemed too great, and hence we decided to make the attempt only with a shadow theatre, and with that idea set at once to work. Two Kasperle comedies of Pocci's were combined, figures were carved from old military post boxes, all their limbs made movable by means of strings and wires, and in the bombers' dug-out the first performance was billed to take place; it was called *All Unnecessary Luggage must go!* And the dream of the shadow theatre was realized. After several weeks winter descended on the land; at the approach of winter the fighting round Riga was over, the advance into Livonia, patrols and outposts in the marshes, and the retreat were abandoned; again the idea of a theatre came into our minds; a few officers of the regimental staff got to know of it and were inspired by our idea, although they believed, as they later told us, that we should fail. We asked for a little support and got to work, animated only by the joy in our hearts; this time we wanted to make a real marionette theatre. "Yes, but who will carve our figures for us? We must go and see Buchner-Konrad of the 10th Company and get him to do that—he is a Munich sculptor. It's to be hoped he doesn't chuck us out when we go to ask him." So over we go to the dug-outs of the 10th Company, and carefully the whole plan is laid before Buchner. Instead of chucking us

FIG. 190. WAGNER, MEPHISTO, AND FAUST
Eastern Front Puppet Theatre of the 2nd Bavarian Infantry Regiment

FIG. 191. GOURTIER, DUCHESS AND DUKE OF PARMA
From the old Faust play
Eastern Front Puppet Theatre of the 2nd Bavarian Infantry Regiment

out he flings himself into the affair with beaming joy, he rushes at once to a place in the dug-out where the clay is got for the ovens, and a couple of days later, when we again visit him, there in the corner which he has chosen as his workshop stand lifelike clay models for Mephisto and Faust. For the old puppet-play of *Doctor Faust* was that with which we desired to start. Now came active weeks. Every free minute we spent in making things—

FIG. 192. FAUST CONJURES UP THE APPARITION OF ALEXANDER THE GREAT
From the old Faust play
Eastern Front Puppet Theatre of the 2nd Bavarian Infantry Regiment

day and night. Buchner's hands, working with the most primitive tools, wrested one figure after another from the elder-wood; these were painted by me and in various other ways embellished. Difficult was it to obtain in the trenches the materials necessary for the costumes, miles away as we were from human habitation. And yet, with some inventiveness, we succeeded. Thus, a woollen helmet became Faust's grey mantle, a handkerchief, which was painted with flaming sulphuret of mercury, was sacrificed for Mephisto's costume. I endeavoured to glue together the whole costume for the lord Orestes from an old sand sack which was then made presentable with noble colourings. Buchner, his patience inexhaustible, made the Duke of Parma and Hanswurst too. Lovely trousers were made of the silk from the parachute of a French rocket light-ball; gilded string was used as an embellishment for the doublets, key-chains as sparkling chains of office; lead shot, matches—all came in handy.

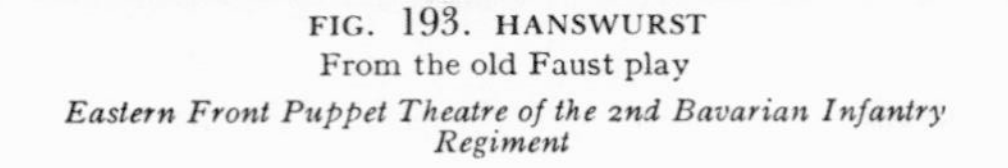

FIG. 193. HANSWURST
From the old Faust play
Eastern Front Puppet Theatre of the 2nd Bavarian Infantry Regiment

FIG. 194. TWO HELLISH SPIRITS
From the old Faust play
Eastern Front Puppet Theatre of the 2nd Bavarian Infantry Regiment

The buttons were indicated by glued-on peas. Unhappily, next morning we found that the rats, much to the annoyance of Buchner, had eaten them. Whereupon I gilded new pea-buttons with bronze strongly smelling of turpentine, and this took away the rats' appetite. The building of the stage itself went on at the same time. We had found a dug-out abandoned because it was in danger of collapsing, and, driven to extremities,

FIG. 195. TWO HELLISH SPIRITS
From the old Faust play
Eastern Front Puppet Theatre of the 2nd Bavarian Infantry Regiment

we set up our common workshop there. Between the half-rotten beams, which formed ceiling and walls, hung down great icicles. An old iron oven which smoked without giving out any warmth gave us the illusion of heat. In addition, we had a miserably functioning carbide lamp, by the irritating light of which the sculptor sat to the right busy carving, and to the left of which the painter was engaged on his backgrounds, wings, and other properties, mostly transmogrified from packing paper and pasteboard; often his colours froze on his brush.

The day of the first performance came terribly near. We had so much to do that we had no time for our meals. How much was still to be rehearsed and learned! Most important of all, the manipulation of the figures. The hours ran on mercilessly toward the first performance. All the officers, including the regimental commander, had promised to attend. The men were curious and, for the most part, as usual, indulged in abuse. "You could do something better than spend our money; your

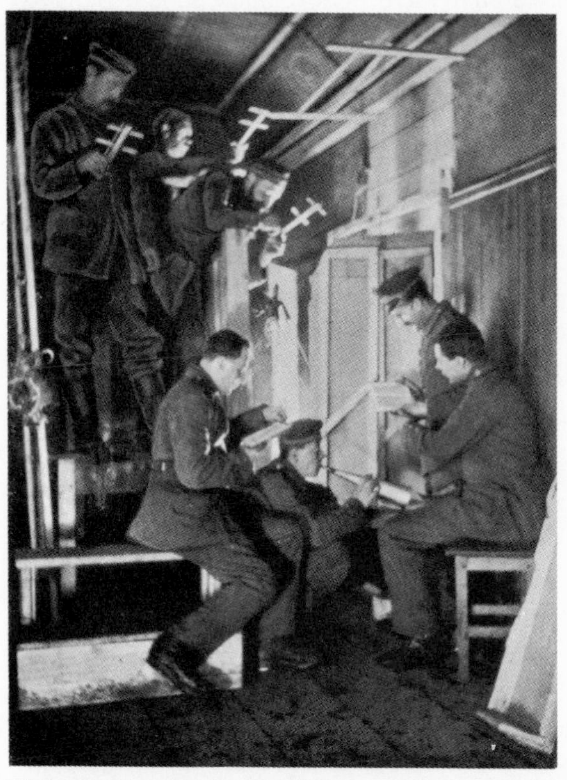

FIG. 196. BEHIND THE SCENES DURING A PERFORMANCE OF A PUPPET-PLAY IN THE 2ND BAVARIAN INFANTRY REGIMENTS ON THE EASTERN FRONT
Eastern Front Puppet Theatre of the 2nd Bavarian Infantry Regiment

wretched theatre has already cost six to seven thousand marks." I was speechless at such reproaches, for the whole thing had as yet not cost the regiment one penny; indeed, we ourselves had paid out of our own pockets a sum of one hundred marks for paint and material. Yet, although they indulged in abuse, they flocked in. The old shed which was transformed into an auditorium was supposed to hold one hundred, but two hundred were crammed in, and four hundred stood outside. The play began, and after the prelude we felt that we had cheered all their hearts by our work. The first scene was listened to with great attention; uninterrupted laughter filled the whole room in the scenes where Hanswurst developed his philosophy of life; and they were deeply touched at Faust's despairing end.

The regimental commander congratulated us on our success, and desired that these performances should be given daily, so that all the men of the regiment might have the opportunity of seeing them. And they came not only from our regiment, for

FIG. 197. THE DOCTOR AND THE THIEF
The former from the *Narrenschneider*, by Hans Sachs, the latter from the *Rossdieb von Fünsing*, by Hans Sachs
Eastern Front Puppet Theatre of the 2nd Bavarian Infantry Regiment

they tramped as long as six hours through deep snow from the adjoining trenches; again and again new crowds filled the house down to the last seat. Hence it came to the knowledge of the divisional staff, the result of which was that one day we were ordered to give a performance with our theatre before the divisional staff and headquarters. Our play made so great an impression that several weeks later we were ordered to give a performance with our theatre at headquarters before the Commander-in-Chief, Prince Leopold of Bavaria. Then came the greatest surprise of all: an order came from headquarters that we had to make an eight weeks' tour with our theatre through the whole province of the Eighth Army in Livonia, Esthonia, and Courland, and produce our plays (we gave, besides our *Faust*, Pocci's comedies and Hans Sachs' farces) in all the towns before the military and the civil population. The success was so extraordinarily great in Dorpat, Pleskau, Narwa, Riga,

FIG. 198. SCENE FROM POCCI'S PUPPET-PLAY "DIE DREI WÜNSCHE"
Eastern Front Puppet Theatre of the 2nd Bavarian Infantry Regiment

FIG. 199. THE THREE PEASANTS
From the *Rossdieb von Fünsing*, by Hans Sachs
Eastern Front Puppet Theatre of the 2nd Bavarian Infantry Regiment

Mitau, and other towns that with our little stage we were able to play to full houses every evening, the pleasant result of which was that we were enabled to contribute quite a nice sum to our regimental funds as the clear profits of our enterprise. Thus our play had a good object, and if, besides, we were successful, among the many thousands who saw our play, in calling forth, here and there, a spark which may perhaps contribute toward keeping the old marionette-play alive, we must have achieved the most ideal object of all.

FIG. 200. FIGURES FROM A FRENCH KASPERLE THEATRE

THE PUPPET-PLAY

OF

DOCTOR FAUST

AN HEROI-COMIC DRAMA IN FOUR ACTS

From the manuscript of the puppet-showman Guido Bonneschky. Published for the first time faithfully in its original form by Wilhelm Hamm, 1850.

DRAMATIS PERSONÆ

FERDINAND, *Duke of Parma*
BIANCA, *his wife*
ORESTES, *his counsellor*
JOHANNES FAUST, *doctor in Wittenberg*
WAGNER, *his famulus*
KASPERLE, *a travelling genius*
MEPHISTOPHELES, AUERHAHN, MEXICO, ALEXO, VITZLIPUTZLI, A GOOD GENIUS — *supernatural spirits*
GOLIATH AND DAVID, THE CHASTE LUCRETIA, SAMSON AND DELILAH, JUDITH, *with the head of Holofernes*, HELENA, *the Trojan beauty* — *apparitions*

ACT I

FAUST'S *study. To the left is a table on which are lying various books and astronomical instruments. In front of the table stands a magic globe.*

SCENE I

FAUST [*sitting alone and reading*]. Varietas delecta . . . "Variety in all things shall create joy and pleasure for man." This is truly a beautiful sentence; I have read it often and often, yet it does not reach far enough for the satisfaction of my desire. One man likes this, another likes that, but we have all the impulse in our hearts to grasp at something higher than we possess. It is true that I might think myself more fortunate than many of my fellows; lacking wealth, lacking support, I have attained by my own efforts to the rank of doctor, and I have carried on this profession honourably for eighteen years. But what is all this to me? Doctor I am, doctor I remain, and beyond that I cannot go in the field of theology. Ha! That is too little for my spirit, which aims at being revered by posterity. I have resolved to apply myself to necromancy, and through that to reach my heart's desire—to make my name immortal.

A VOICE TO THE RIGHT [*invisible*]. Faust! Faust! Leave off this project! Pursue the study of theology and you will be the happiest of men!

A VOICE TO THE LEFT. Faust! Leave off the study of theology! Take up the study of necromancy and you will be the happiest of men!

FAUST. Heaven! What is that? To my intense amazement I hear two invisible voices! One on the right warns me to keep to the study of theology, and says that if I do so I shall be the happiest of men; that on the left advises me to take up the study of necromancy, and says that if I do so I shall be the happiest of men. Well, then, I shall follow you, you voice to the left!

THE VOICE TO THE RIGHT. Woe, O Faust, to your miserable soul! Ha! Then you are lost!

THE VOICE TO THE LEFT [*laughs*]. Ha! ha! ha! What a jest!

FAUST. Again do I hear these two voices, one on the right bewailing and one on the left laughing at me? Yet must I not alter my purpose, since I feel that only through the study of necromancy can I bring my desires to satisfaction! Yet again, I shall follow you, you voice to the left!

SCENE II

FAUST *and* WAGNER.

WAGNER. I come to inform your Magnificence that two students gave me, to be handed to your Magnificence, a book which you have long wished to possess, since it deals with the study of necromancy.

FAUST. Since yesterday I have been hoping you would bring it me, and I have been eagerly awaiting it. Where is the book?

WAGNER. I have laid it in your lecture-room.

FAUST. Good! Leave me!

WAGNER. In all humility, your Magnificence, I should like respectfully to ask about something that worries me.

FAUST. Speak on, my dear Wagner! You know that I have never refused you anything that lay in my power to give.

WAGNER. Your Magnificence, I would humbly beg you to let me engage a young lad to be taken into your service, for it will be too difficult for me to manage the household and at the same time carry on my studies.

FAUST. Do so, my dear Wagner! It has long been my wish to see you less burdened by the management of the house so that you can apply yourself more freely to your study. You can therefore look round for a young lad, and when you have found a fit person, who can show you good testimonials, you may bring him to me, and I shall make further arrangements with him.

WAGNER. Very good, your Magnificence! I shall do all in my power to fulfil your commands.

FAUST. Then the book that the students handed over to you is in the lecture-room?

WAGNER. Yes, your Magnificence!

FAUST. Then I will go and see whether it is the same that I have vainly sought to get for so long. [*Goes out.*

SCENE III

WAGNER [*alone*]. What a noble-minded master is this! What would have become of me without his help? He has taken me, a lonely orphan, into his house, and has always taken such pains over my education and my progress. Can I ever repay him for all his kindness? I think not. But I will work honestly, do all I can to anticipate his wishes, and at least show him that he did not expend his kindnesses in vain, that a thankful heart beats in my bosom for him. I shall therefore go to a good friend of mine and inquire whether perhaps he knows of a suitable young man to enter his service. Accomplishment of a duty is the greatest proof for a man who wishes to show his gratefulness. [*Goes out.*

SCENE IV

FAUST [*who enters a little earlier with a book in his hand*]. Good man! Just remain as I have found you hitherto and I will truly strive to recompense you for your trust, and to further your fortunes so far as I can. Now, however, I desire also to carry out my resolution to devote myself solely to the study of necromancy by means of this book, and I desire immediately to test this art by a conjuration. "I charge you, you furies of hell, by hell's gates, by Styx and Acheron, to appear immediately before me! Break out, you howling storm, that Ixion's wheel may stop and Prometheus' vulture forget to torment him, and carry out my will! Despair, Fury, and Rage—hurl you at my feet!"

SCENE V

FAUST, ALEKSO, VITZLIPUTZLI, AUERHAHN, MEXICO [*hurled in with wild thunder and lightning*]. *Later* MEPHISTOPHELES.

FAUST. A pretty crew! Yet you are very negligent! Don't do that again to me! Tell me, you first fury of hell, what's your name?

MEXICO. Mexico.

FAUST. And how quick are you?

MEXICO. As quick as a bullet fired from a gun.

FAUST. You have a great speed, but not enough for me. Vanish!

[MEXICO *disappears through the sky.*

FAUST. And what name have you, hellish fury, and how quick are you?

AUERHAHN. Auerhahn, and I am as quick as the wind.

FAUST. That's very quick, but not enough for me. Vanish!

[AUERHAHN *disappears in the same way as* MEXICO *did.*

FAUST. What's your name, hellish spirit?

VITZLIPUTZLI. Vitzliputzli.

FAUST. And how quick are you?

VITZLIPUTZLI. As quick as a ship sailing on the sea.

FAUST. As a ship sailing on the sea? That is a fair speed, but with unfavourable winds it doesn't always reach its goal. Vanish!

[VITZLIPUTZLI *goes off.*

FAUST. Say on, hellish fury—what's your name?

ALEKSO. Alekso.

FAUST. Alekso? And what's your speed?

ALEKSO. I am as quick as a snail.

FAUST. As a snail? So you are truly the slowest of all the hellish spirits? I can't make any use of you at all. Vanish!

[ALEKSO *goes slowly off.*

FAUST. Tolerably slow!

[MEPHISTOPHELES *comes in dressed like a hunter.*

FAUST. Ha! What do I see? A hellish fury in the likeness of a man?

MEPHISTOPHELES. You must know, Faust, that I am a prince of hell, and have the power to assume, and to appear in, whatsoever shape I please.

FAUST. You a prince of hell? What's your name?

MEPHISTOPHELES. Mephistopheles.

FAUST. Mephistopheles? That's a good-sounding name. And how quick are you?

MEPHISTOPHELES. As quick as man's thoughts.

FAUST. As man's thoughts? Ha! That is an extraordinary speed; for I can be with my thoughts one moment in Africa, another in America. Say, hellish fury, if you wish to serve me. I shall promise you to be yours, body and soul, at the end of the time I shall settle on.

MEPHISTOPHELES. Tell me the conditions, Faust, to which I must submit.

FAUST. The first condition is this, that you get me as much money as I ask from you. The second is that you make me a person of consequence among all great men and at all great courts, that you carry me wherever my desire takes me, that you warn me of all dangers. And the third condition is that you tell me before our contract comes to an end and that you obey me for four and twenty years. Are you willing?

MEPHISTOPHELES. Why four and twenty years, Faust? Half of that is enough.

FAUST. Four and twenty years—no less.

MEPHISTOPHELES. Well, I am satisfied. [*Aside*] I'll cheat him all right; he doesn't count on my skill. [*Aloud*] But, Faust, I must leave you just now to tell my prince Pluto of our agreement and ask him whether I may conclude this contract with you.

FAUST. Leave me then! But when will you return to me?

MEPHISTOPHELES. As soon as you think of me, Faust. [*Goes off.*

FAUST [*alone*]. As soon as I think of you? I ought not certainly to enter into an agreement with a hellish spirit, but it is the only way to accomplish my desires quickly. And have I not sufficient power through my knowledge to get out of his clutches when half the time of the contract is over? Yes, so be it! Yet I feel so weary, so exhausted; this is certainly the result of the exertion of my spirit. I will go into my cabinet to rest for an hour or so, and then I shall carry on my plan with renewed energy. [*Goes out.*

SCENE VI

CASPER [*comes in with a bundle on his back*]. Pox on't! There's travelling for you! I have walked fourteen miles in thirteen days, and only every half-hour I've struck an inn! I've come to-day from—here. Here in my knapsack I have my whole equipage: I have the lining for a new overcoat—the cloth for it, however, is still with the shopkeeper. And here a half-dozen good stockings without heels, and a whole dozen of shirts. The only trouble is that the best of them has got no sleeve, and the eleventh is patched with the twelfth. I bought myself this fine beaver in Leipzig. It cost me twenty-one groschen, and a pair of new turned-up shoes of the latest style, the heels tipped with nails, cost me likewise seventeen groschen six pfennigs. Yes, yes, travelling costs money; I note that my purse has fallen into a galloping consumption. I've been journeying in Holland, Scotland, Brabant, England—but I've got to get there first. But in Danzig, Breslau, Vienna, Regensburg, Friessland, Nürnberg, Dresden—I've never been there at all. I was even three hundred miles behind the New World, but all at once I came on a wall and couldn't go on any farther; then I turned back, and now fortunately I find myself in Wittenberg and shall see whether I can get a situation, for I'm fed up with wandering about. When my father took leave of me said he: "Casper, try only to set your affairs a-swinging." That I've done, for my bundle is so light I could chuck it over the biggest house. But zounds! What sort of an inn is this, where neither landlord nor waiter is to be seen! Hullo, there! Wake up, household! There's a new customer here who wants to get a two-groschen bottle of wine! Mr Landlord! Waiter! What the devil's become of you! Aha! Now I hear some one coming. I'll give him a good fright! [*He creeps under the table.*

Scene VII

Casper *and* Wagner.

Wagner. I'd like to know how it is that I can't find anybody who wants to go into service. The friends I was with are quite unable to recommend a suitable person to me.

[Casper *comes out and frightens him.*

Wagner. Oh, heavens! What is it? A strange man in my master's room?

Casper. You're trembling like an anvil.

Wagner. Who are you, my friend? How dare you come into this room without getting yourself announced? How did you get in?

Casper. What a comic question! I came in on my feet, of course! But tell me—is it the habit and practice in Wittenberg for a customer to get himself announced when he wants to buy a bottle of wine?

Wagner. My friend, you're not in an inn.

Casper. No?

Wagner. No, on the contrary, you are standing in the study of his Magnificence Doctor Faust.

Casper. Well, well! What mistakes one can make! I thought, seeing so many young men go into this house here, that it must be an inn, for there also many folks are constantly going in and out.

Wagner. That doesn't follow, as you shall hear, for all the young men you saw are students who come daily to the lectures which are delivered in this house.

Casper. Students? And I thought they were customers! Then I made a pretty mistake! Well, I'll set myself on my feet to find an inn. [*Makes as if to go off.*

Wagner. Wait a minute, my friend! To judge by your clothes you are a servant?

Casper. Yes, I've got my master on my back.

Wagner. And where have you come from now?

Casper. From Italy.

Wagner. Yes, my friend, but Italy is big. What I wanted to know was the precise place of your birth.

Casper. Oh! You wanted to know the place of my birth?

Wagner. Yes.

Casper. I was born in Calabria.

Wagner. What? In Calabria?

Casper. Yes, in Calabria.

Wagner. And what made you leave it?

CASPER. There I was engaged as companion in the house of a study-maker.

WAGNER. In the house of a student I suppose you mean?

CASPER. No, in the house of a study-maker! Surely I know best in whose house I was?

WAGNER. So you were given a situation in the house of a student?

CASPER [*aside*]. Oh, Jeminy! Isn't that a stupid fool! But I'll let him be. [*Aloud*] Yes, in the house of a student.

WAGNER. And what made you give up this situation?

CASPER. Well, do you see, that was a curious story. Every morning in Calabria I had to bring into the college for my former master his copy of Donatus, a very big book, and every day had to go over a certain plank. Now, one morning in the middle of this plank I met a pretty girl; I made her a couple of compliments, missed my footing, and let the book fall into the water. When my master heard of this he chucked me out of the job.

WAGNER. And then you started to travel?

CASPER. Yes, *per petes apostolorum*.

WAGNER. You've got some kind of a testimonial?

CASPER. Oh, yes! My master wrote a testimonial for me in black letter, chancery hand, and current script so naturally on my back with a gnarled stick that the letters will be legible till domesday.

WAGNER. Do you want to take up service again?

CASPER. Oh, yes! Do you want a servant?

WAGNER. Yes, if only you had good testimonials to show.

CASPER. Well, I've got them on my back.

WAGNER. Yes, sir, but he whose famulus I am will not be satisfied with that.

CASPER. Who are you?

WAGNER. Famulus to Doctor Faust.

CASPER. You are his famulus and want to engage me? Then I'd be a servant's servant?

WAGNER. No, you don't understand me; I will explain. My master has asked me to find a young fellow who wishes to go into service.

CASPER. Oh, that is a different story.

WAGNER. If you desire to enter his service you need only say so.

CASPER. Of course! But listen—there isn't too much work to be done here, is there? Because, do you see, I'm no great friend of work. I have so long a finger on each hand that I'm always knocking myself.

WAGNER. Oh, you'll get a very good place, with good food and wages, for my master is not married. Only one thing I must advise you about—keep secrets!

CASPER. Don't you worry about that! Just you wrap up the secrets I must keep in a piece of roast beef and a bottle of wine, for as long as my mouth has something to chew and my throat to swallow I'm as dumb as a fish.

WAGNER. Well, that will be all right. I can also tell you at once what duties you have. In the morning you must remove the dust from the *repositorium* and the books.

CASPER. Remove the books from the *repositorium* and the dust? Oh, that I can do; I've often done that in the house of my former master.

WAGNER. Then chop wood and draw water; that's all you have to do.

CASPER. Chop wood and draw water? Oh, the devil!

WAGNER. Don't worry about that! You'll certainly like working at my master's. And moreover, if only you are trustworthy and industrious, and when we're better known to one another, then I'll do still more for you, for I have the key of the wine-cellar in my keeping.

CASPER. See here, Mr Famulus, how would it be if we exchanged places? You would give me the key of the wine-cellar and you could see to the wood-chopping and the water-drawing.

WAGNER [*laughs*]. Ha! ha! ha! That wouldn't be bad, but, as I said just now, all that will come in time.

CASPER. Yes, yes, all that will come. But couldn't you give me something to eat and drink just now—a leg of mutton, say, or a pheasant and a bottle of wine—for I've come from a journey and have such a desperate hunger that my stomach is as shrunk up as an empty tobacco-pouch?

WAGNER. Follow me! I'll fulfil your wish directly. [*Goes.*

CASPER [*calls to him*]. Mr Famulus! Mr Famulus!

WAGNER [*comes back*]. Well, what do you want?

CASPER. Just show me a spot where I may put away my wardrobe, so that once and for all I may take the hump out of my bundle.

WAGNER. Just come with me into another room. There you can put away your clothes and can apply yourself to eating and drinking. [*Goes out.*

CASPER. Rejoice, belly; now a treat's in store for you! He! he! Mr Famulus, just take me with you! I don't know my way about this house. [*Goes out.*

SCENE VIII

CASPER [*runs in*]. Zounds once more! I've been in the kitchen and have inspected what there is to eat—bacon frizzling loud and wine from a pump! The boot-polish is good for nothing. Well, I'm glad that I have got a new situation and that a good one. I have taken off my knapsack and wish to look round. Zounds once more! Where has the plum jam been put? This is a queer house, with all the rats' tails and the piles of books, which are as big as my grandmother's bread-board. Zounds! What is there? Is that truly a tailor's measure? Am I then to serve a tailor? [*He steps on to the magic circle which is traced on the ground and turns over the pages of the books that lie on the table.*] Zounds once more! A tailor hasn't so many books. I can read them too. What's this?

Brrrr! What's this? *Brrr!* [*Shakes his head vigorously.*] That is a K—Katz—D, B, U, B—Pudel—Katzpudel; K–E–K——Karek Barek, B–E–R——Berlicke! [*Three infernal spirits appear.*] Berlicke! Berlicke! [*He looks round.*] Oh! Lord Jesus! Lord Jesus! Help! Help! What do you black fellows want? There's no chimney to clean here. What do you want, you charcoal-burner with the red nose? Oh, dear god! Dear god! Zounds once more!

THE SPIRITS. Just come out of that circle, and we shall tell you. We await your commands.

CASPER. No! no! I won't come out of this because you ask me to! Zounds once more!

THE SPIRITS. But you must step out and give us your hand, or we do not go away.

CASPER. Come out? Ne—give you my hand? Ne—then you'll go, you dirty chimney-sweeps? No, I'm not coming out! I'm not coming out! Stay as long as you will! Who asked you to come here?

THE SPIRITS. You yourself summoned us through saying Berlicke.

CASPER. Then I'll just summon you off again. How am I to do it? Zounds once more!

THE SPIRITS. That you must do through saying Berlocke.

CASPER. Aha! Spiritus, do you mark that? You wait, you rats' tails! I will hunt you down now! So, now watch! Berlocke! [*They vanish.*] Berlicke! [*They come again.*] Berlocke! Berlicke! Berlocke! Berlicke! Berlocke! Berlicke! Berlocke! Berlicke!

[*He calls out ever quicker and quicker, the infernal spirits come and vanish ever more swiftly, until at last they fling Casper over the houses with woeful shrieks.*

ACT II

The same room as in Act I. The chair stands on the left.

SCENE I

FAUST [*enters*]. Oh, heavens! What a strange dream has disturbed me to-night and thrown my soul into torment! I thought I saw an angel who warned me to abandon my project of entering into an agreement with the hellish spirits, else now and hereafter I should be utterly lost. But I cannot bring myself to change my plan, for I feel that I am cunning enough to cheat Satan, with all his craft and his tricks, and to cancel the bond itself whenever my wishes are fulfilled. That vision was probably but a figment of my troubled imagination, a phantom of the mind, designed to frighten me as I stretched out my hand for this treasure. But the ancients say, if one grasps it fearlessly it will vanish completely. So I will if Mephistopheles . . .

Scene II

Mephistopheles *enters*.

Mephistopheles. Well, Faust, have I kept my word?

Faust. To my greatest astonishment. Have you got permission from your Pluto to serve me?

Mephistopheles. Yes, Faust. But he requires a deed made out in your writing, saying that you will be his property, body and soul, at the expiration of the time fixed by you.

Faust [*goes to the table*]. I will fulfil his desires immediately.

Mephistopheles. What are you going to do, Faust?

Faust. Sign my name.

Mephistopheles. Among us of the Plutonic realm no signature with ink is valid. Among us it must be written in blood.

Faust. But how can I get blood without cutting one of my limbs and so giving pain to myself?

Mephistopheles. Put your hand to my mouth and I will provide you with some blood painlessly.

Faust [*gives him his hand*]. Here! [Mephistopheles *blows on it.*

Faust. In truth, my blood flows without my feeling anything at all, and to my amazement it comes forth in two letters—an H and an F. What do these two letters mean, Mephistopheles?

Mephistopheles. What, Faust? Can't you, who are so great a scholar, interpret these letters? Well, then, these signify *Homo, fuga*, or "Man, flee."

Faust. Ha! From whom should I fly, infernal spirit?

Mephistopheles. You must not interpret it as a bad omen—it means fly into the arms of your true servant Mephistopheles.

Faust. If that is so I shall, without other thought, sign the document with it. [*He writes.*] "Johannes Faust"! So, now you can carry this contract to your prince Pluto.

Mephistopheles. No, Faust; henceforth I do not move a step from you. Tell me, who should carry off this document—a wolf, a bear, or a tiger?

Faust. What should such fierce beasts be doing in my room? Let a crow take it.

Mephistopheles [*nods*]. Watch, Faust!

[*Thunder and lightning. A crow appears, takes the bond in its mouth, and flies off.*

Faust. But will the crow deliver the bond correctly?

Mephistopheles. You can rest assured of that. Have you now any commands for me, so that I can show you how quickly I can fulfil them?

Faust. No. Withdraw now until I call you.

Mephistopheles. Very well, my Faust, only mention my name and in the twinkling of an eye I shall be with you. [*Goes off.*

Faust [*alone*]. Now I should only like to know where my famulus remains so long. He could have done my bidding and been back ages ago. I hope no misfortune has befallen him; his long absence makes me very anxious. Ah! There he comes at last!

Scene III

Wagner *enters.*

Wagner. I inform your Magnificence that I have done everything you commanded me.

Faust. Good, my dear Wagner. But how comes it then that you have not brought back a young lad as I gave you leave and as you yourself wished? Perhaps you have not been able to find anybody suitable?

Wagner. Oh, yes, your Magnificence, when you bid me I'll bring him to you at once.

Faust. Good! Bring him in! Where is he?

Wagner. In my room. He has just come from a journey and has asked me to give him something to eat and drink. I will call him at once. [*Goes to the side.*] My friend!

Casper [*from without*]. What is it?

Wagner. Come here! His Magnificence wants to speak to you.

Casper. In a minute! Let me just eat up my leg of mutton!

Wagner. His Magnificence can't wait for you. Come at once!

Casper. Just let me drink down this little glass of wine!

Wagner. Don't be so rude! Come at once!

Casper. What are you shouting about? I'm ready and waiting all the time.

Scene IV

Casper *enters.*

Casper. Well, what's the matter that you make such a fuss that one can't eat in quiet?

Wagner. Here! His Magnificence wants to speak to you.

Casper. Ah! That's quite a different matter—I'll make my compliments to him at once. Your Insolence, it gives me uncommon pleasure that you have the honour of making my acquaintance.

Faust. Have you been in service before?

Casper. Yes, your Insolence, in Calabria, where I was in the service of a study-maker.

Faust. Are your parents still alive, my friend?

Casper. I'm not quite sure. I always think that the drum-pigeons hatched me.

Faust. You have good testimonials to show, I suppose?

Casper. Rest assured; I have a most magnificent testimonial—[*aside*] on my back.

Wagner. Yes, your Magnificence, he has assured me of it [*speaking confidentially to* Faust].

Casper [*sits on the chair*]. Ah! Here's a chair; I can make myself comfortable. [*Sits down.*] Zounds! What a charming seat! This chair must be upholstered with fat steel springs! *Prr! Prr!* Here one can give oneself an air of authority.

WAGNER. My friend!

CASPER. Well, what is it?

WAGNER. Get up at once!

CASPER. Why?

WAGNER. It isn't proper to sit down in the presence of his Magnificence.

CASPER. Oh, that doesn't apply to me! I've come from a journey, and he who comes from a journey is tired and desires to be comfortable.

WAGNER. But this chair is his Magnificence's own!

CASPER. Just now it is Casper's own! I don't know at all why Mr Famulus finds fault with it. Your Insolence hasn't said a single word, but every moment you, Mr Famulus, find something that doesn't seem right. I say this to you: if this goes on and once I get up my temper, then . . . [*At this moment the chair on which he sits is enveloped in flame.*] Oh! Oh! Help! Help! [*Goes out.*

SCENE V

MEPHISTOPHELES, CASPER *singing*.

CASPER [*joining* MEPHISTOPHELES]. Well, who are you, *mon cher ami*? Is it good manners to come into his Insolence's room with your hat on your head?

MEPHISTOPHELES. Do you not recognize me then? I am the master's huntsman.

CASPER. The master's huntsman? And what does master want a huntsman for? He's a theologian.

MEPHISTOPHELES. But a very great lover of the hunt. I stand very high in his favour, for I catch foxes and hares with my hands.

CASPER. Zounds! Then you're a clever fellow! You know how to spare powder! What's your name?

MEPHISTOPHELES. Mephistopheles.

CASPER. What? Stoffelfuss, did you say?

MEPHISTOPHELES. Mephistopheles! Don't mutilate my name, or——

CASPER. Well, well, don't shout in that way and take on so! I didn't get it right.

MEPHISTOPHELES. Have you heard that our master is going to travel?

CASPER. Travel? Where to?

MEPHISTOPHELES. Just travel.

CASPER. Just so, travel. Where to?

MEPHISTOPHELES. To Parma.

CASPER. Oh, you can't make a fool of me! What would he want to get into a *Barme*[1] for, when he's so fat?

[1] Manger.

MEPHISTOPHELES. No! no! Parma is a principality where a great nuptial ceremony is to take place.

CASPER. Oh! That's quite another matter. But what's a nuptial ceremony?

MEPHISTOPHELES. A nuptial ceremony is a marriage.

CASPER. Do they give you much to eat and drink there?

MEPHISTOPHELES. Oh, yes: lots.

CASPER. Rejoice, my belly: there's going to be another downpour! Is master to take me with him or not?

MEPHISTOPHELES. No, he has ordered that you should remain behind. He is to journey quite alone to Parma on his robe.

CASPER. On his robe? That will be a pretty wear and tear.

MEPHISTOPHELES. Yet I will take it on my own responsibility to take you along with us, without his being aware of it.

CASPER. Oh, yes, do that, good Stoffelfuss, for I simply adore eating and drinking!

MEPHISTOPHELES. Would you prefer to ride or go in a carriage?

CASPER. You know best. Get something sent here for me to ride.

MEPHISTOPHELES. Very well, I'll see to it at once. But I expressly forbid you to tell anyone in Parma who our master is or what his name is—otherwise I'll break your neck. Do you understand me? [*Goes out.*

CASPER [*alone*]. Oh, yes, I've understood all right. One would need cotton-wool in one's ears not to have understood. Well, I must go and get my equipage packed up. And he will get me something to ride on; if only he brought me here a nice little Polish or Hungarian pony, for I like riding, and—[*A dragon enters and gives him a knock on the shoulders; he falls down.*] What kind of behaviour is that? [*Gets up.*] He! he! Help! help! Stoffelfuss! Stoffelfuss! Is that the horse you promised me?

MEPHISTOPHELES [*coming in*]. Just get up on it. It won't harm you.

CASPER. Oh, yes, just get up on it! That's a new-fangled horse. I must think it over a bit. Zounds! It's got a walking-stick behind! If the animal hits me a knock on the nose it'll upset the applecart. It's even got wings! Won't I suffer if it flies off? Courage! I'll get up. [*He mounts.*] That's not a bad seat! Hi! hi! Little fox! Hi! [*The dragon gives him a knock at the back of his head with its tail.*] Well! And who's the lout who's given me a blow as if he wanted to knock out my four senses? I'll put a stop to that! Well, little fox, gee up, gee up! Hi! hi! [*The dragon gives him another blow.*] Thunder and lightning! Some one has struck me again! [*He turns round.*] I believe it was you with your walking-stick! Just you wait! I'll take another seat so as to get out of the line of fire. [*He sits farther forward.*] This beast's contrived not badly for a learner in riding, for if one were to fall off one couldn't fall far. [*Makes movements as if he were walking.*] But I hope there won't be bad weather to-day; I fear this journey will cost me more in shoe leather than I get in salary.

Now, little fox. Hi! hi! [*The dragon rises suddenly, and flies upward amid thunder and lightning. He cries out.*] He! he! Help! help! Stoffelfuss! Stoffelfuss! The animal's going up into the sky! He! he! he! [*He vanishes.*

The curtain falls

ACT III

SCENE I: *A garden.*

DUKE, BIANCA, ORESTES.

DUKE. Well, dearest spouse, how do you like my Court? Can you find in me and my subjects sufficient recompense for the sacrifice you have made in leaving your parents and your native country?

BIANCA. Oh, my husband, how deeply you shame me in asking that question! Have I not received proofs sufficient of your love and of the esteem of your subjects to make me happy in the thought of spending my life at your side?

DUKE. And yet it seems to me as if a secret grief sometimes clouded your face. Have you perhaps a secret in your breast in which you do not feel perhaps that I am worthy to participate?

BIANCA. No, I am only troubled by sorrow for my father, who was seriously ill when I accompanied the ambassadors here to you. Feeble and exhausted, he raised himself in his bed when I took leave of him and said: "Be as good a wife as you have been a daughter and my blessing will follow you always." Oh, my husband, from that time I have been unable to repress the anxiety which grief for my father's health occasions me. Will you grant me a request?

DUKE. Speak, dear Bianca—it shall be granted.

BIANCA. Allow me to leave you now, so that I can give the ambassadors some consoling words to carry to my father, telling him of the love and kindness which I have met with here.

DUKE. With pleasure I accede to your wish. Only I beg you not to withdraw your gracious presence overlong.

BIANCA. I will hasten as quickly as I can and come to you again. [*Goes off.*

SCENE II

DUKE *and* ORESTES.

DUKE. Well, dear Orestes, what do you say to my choice of a wife?

ORESTES. That you must consider yourself and the whole country lucky to have gained so virtuous a consort and duchess, and may Heaven grant the happiness to us all to be ruled right long by such an excellent pair of princes!

DUKE. I thank you for your wish, and hope that you will aid me even further with your wise counsel to make my subjects happy.

ORESTES. Truly, your Highness, I will try to show myself worthy of the great favour you display towards me.

DUKE. Have you prepared, as well as possible, everything that will be needed for these nuptials?

ORESTES. Yes, your Highness. I have not failed to carry out your orders to the best of my ability. I have also got it announced by means of a proclamation that all artists and scholars should come here to embellish this marriage by their presence.

[CASPER *is hurled down from the sky on to the stage.*

DUKE. Good heavens! What's that?

ORESTES. Your Highness, I am amazed. This man——

DUKE. Ask him, dear Orestes, how it can possibly be that he could come down here out of the sky?

CASPER. Him! ham! hum!

ORESTES. My friend, who are you?

CASPER. Him! ham! hum!

ORESTES. How can it be that you have fallen down here out of the sky without injuring yourself?

CASPER. Him! ham! hum! Him! ham! hum!

DUKE. He seems to be dumb, if he's not dissembling. Promise him twenty ducats if he reveals this secret to us; if not, he'll get twenty good lashes.

CASPER. Him! ham! hum! Bum! bum!

ORESTES. My friend, you have heard what his Highness has promised you. If you can speak, don't delay any longer in fulfilling his Highness's desire.

CASPER. Him! ham! hum! Him! ham! hum!

DUKE. I truly believe that this man is merely laughing at us. Orestes, call the watch at once!

ORESTES. In a moment, your Highness. [*Prepares to go off.*

CASPER [*holds him back*]. No! no! Just wait, old sir!

DUKE. What? You can speak? You're not dumb?

CASPER. That's just the trick.

DUKE. You seem to me to be a very obstinate fellow.

CASPER. Oh, no. I'm a very good man, but I have a very bad companion, and he has forbidden me to speak.

DUKE. What's your name?

CASPER. Ah! That's what I must not tell.

DUKE. Then you must have been guilty of some terrible crime if you dare not reveal your name.

CASPER. You don't think, do you, that I've stolen anything? God forbid! Casper can go wherever you will, but no one can say that about Casper.

DUKE. So Casper's your name?

CASPER. Who told you that I was called Casper?

DUKE. You yourself.

CASPER. I? [*To himself*] Oh, you damned blubberer!

DUKE. To judge by your dress you are a servant?

CASPER. You've guessed it.
DUKE. What's your master's name?
CASPER. Ah, there's the rub! I must not reveal it, else my neck'll be broken.
DUKE. You can always tell it to me. No harm shall come to you.
CASPER. Who, then, are you?
DUKE. I am the Duke of Parma.
CASPER. What? The Duke of Parma? Pray pardon me for not having yet made you my compliments. I am very glad that you have the honour of my acquaintance. Go on!
DUKE. Very good, my friend. Well, I should gladly know the name of your master.
CASPER. I must not tell it; but I'll show it you pantomimically.
DUKE. Well, I am content.
CASPER [*raises his arm*]. See here! What is that?
DUKE. That's an arm.
CASPER. Well, what is this just in front? Just in front?
DUKE. A hand, and if you close it it's a *Faust*.[1]
CASPER. Right—I serve him. But I didn't tell you.
DUKE. What? The great scholar? You're in Doctor Faust's service?
CASPER. Yes, I am. It's great to be able to speak to people and yet not give away secrets. That's the point—I am a true genius; I've always got a bunch of tricks in my head.
DUKE. Since you live with this famous man, have you not learned some of his art?
CASPER. My master learned everything from me.
DUKE. From you?
CASPER. Oh, yes! I am Faust's teacher. Haven't you heard, then, of my skill? My name has been blazoned abroad to all four corners of the earth.
DUKE. No, I haven't heard anything about it.
CASPER. Aha! Now I call to mind, we had so hard a winter then that all the sounds were frozen; but just let it thaw, and my fame will make a devil of a row.
DUKE. I should be glad to see some of your art.
CASPER. So you want to see some of my art?
DUKE. Yes!
CASPER. You shall have it directly. [*Aside*] See, Casper, if you had learned anything now, how you could have profited. [*Aloud*] Do you wish to see something big, something grand?
DUKE. Yes, something extraordinary.
CASPER. Would you like to see a great big wave come rolling in to drown all three of us?
DUKE. No, I shouldn't like to see that.
CASPER. That's a very big piece.
DUKE. I'd rather you showed me something else.

[1] Fist.

CASPER. Something else? Perhaps you'd like to see a great millstone crashing down from the sky to beat us down ten fathoms deep in the bosom of the earth? That's a very impressive piece.

DUKE. No, I shouldn't like to see that; my life would be endangered in this as in your first piece. Something fine, something pleasant —that's what I want.

CASPER. Oh! Something fine? Perhaps you'd like to see an Egyptian darkness wrapped up in cotton-wool? That's a very fine piece. But I need four weeks' time to pack it in its box.

DUKE. Don't go on talking such utter nonsense!

CASPER. Utter nonsense? Can you do it then?

DUKE. No—but——

CASPER. Well, you mustn't say that it is utter nonsense, for I can become offended so quickly as to make my body all run into gall when a person steps up and says it is utter nonsense, and yet can't do it himself.

DUKE. Don't be so indignant about it! Show me something else.

CASPER. Perhaps you'd like to see a devilry?

DUKE. Yes, I should like to see a *Salto mortale*.

CASPER. But what's my reward if I do show a devilry?

DUKE. I have laid aside twenty ducats for you, and these you shall get.

CASPER. I should be right glad if you gave me the money in advance.

DUKE. Why? You don't doubt my promise, do you?

CASPER. Oh! God forbid! But, do you see, when I make a devilry, usually I stay three or four months in the sky, so it would be an advantage if I had the money in advance wherewith, among other things, to pay for my lodgings.

DUKE. So soon as I am convinced of your skill you shall have the ducats—but not before.

CASPER. So I get no money in advance?

DUKE. No.

CASPER. Well, for my part, if an accident happens to me I'll have no gold with me, and you shall have it on your conscience.

DUKE. Yes! yes! Just give me a proof of your skill.

CASPER. At once! At once! Just stand a little to the side, and I will start my invocation. [*He turns always on one foot.*] *Br! br! br!*

DUKE. What does that mean?

CASPER. That's the invocation. You mustn't put me out. *Br! br! br!* Well, sirs, shut your eyes, in case things spring into your face.

DUKE. What kind of things?

CASPER. Sugar and coffee! *Br! br! br!* What do you really want to see, sirs?

DUKE. Well, a caper!

CASPER. Make it yourself! I can't do it! [*Goes off.*

DUKE. Wait, you damned rascal! Orestes, go and get the watch to arrest him and let him suffer for his villainy.

ORESTES. At once, your Highness. [*Goes off.*

Scene III

Duke *alone.* *Then* Faust. *Later little by little the apparitions.*

Duke. I will not let him be punished too severely for his audacity in having ridiculed me, for he has given me real pleasure with his droll conceits. If I'm not mistaken, a man whom I don't know is coming down the alley towards me.

Faust [*enters*]. Most serene Duke, with deepest humility I beg you to pardon me for making so free a visit. But as it has been proclaimed that all artists and scholars are invited to come to your princely marriage, I have hastened to fulfil your Highness's commands, and humbly beg that you will permit me to present here my art and my skill.

Duke. What is your name?

Faust. Johannes Faust.

Duke. What? You are the world-famous Doctor Faust, whom all men admire, you who are able in one minute to summon summer and winter, like nature itself? You are very welcome to my Court: for long I have desired to make your personal acquaintance.

Faust. Your Highness overwhelms me with your praise, which up to now I have not done anything to deserve. Perhaps your Highness would desire to see some proofs of my art?

Duke. If it gives you no trouble, I should accept your offer with pleasure.

Faust. At all times and in every place, your Highness, I am ready to fulfil your commands.

Duke. Well, then, I should like to see here the big giant Goliath and the little David.

Faust. Your Highness shall immediately be satisfied! Mephistopheles! Do you hear? Cause the giant Goliath and the little David to appear immediately!

[*An adagio sounds, and* Goliath *and* David *appear. The latter has his sling in his hand. After some minutes, during which the* Duke *speaks to* Faust, *the* Duke *indicates that he has seen enough.* Faust *bows and nods*; *the apparitions vanish; and the music ceases.*

Duke. Indeed, you have shown me these two persons beyond my expectation. I have long wondered how it was possible that this giant could be killed by a sling wielded by so small a man.

Faust. It is a proof that the strong must not always trust to their strength. What would your Highness like to see further?

Duke. The chaste Lucrece, as she stabs herself in the breast on the Capitol at Rome, since her virginity was in danger of being violated by force.

Faust. Very well, your Highness! Do you hear, Mephistopheles? Cause the chaste Lucrece to appear!

[*He nods. Adagio.* Lucrece *appears with the dagger at her breast. Then the same dumb show as before.*

DUKE. Upon my honour! Most excellently do you gratify my wishes. This woman through her chastity redeemed Rome's tottering imperial throne. She will shine in history as the greatest example of womanly virtue.

FAUST. Your Highness is right! What would your Highness desire to see now?

DUKE. Samson and Delilah, as she was cutting off his hair.

FAUST. In a moment, your Highness! Mephistopheles, cause Samson and Delilah to appear!

[*He nods. Adagio. The curtain at the back opens and reveals* DELILAH *sitting in a chair with scissors in her hand; before her sits* SAMSON, *sleeping with his head laid on her bosom; she is about to cut off his hair. Then dumb show as above.*

DUKE. You are giving me more and more proof that you are one of the greatest magicians of our time. But I should like to ask for one thing more.

FAUST. Make your command, your Highness!

DUKE. I should like to see the heroic Judith with the head of Holofernes.

FAUST. Very well, your Highness! Mephistopheles! Do you hear? Cause her to appear at once—Judith with the head of Holofernes!

[*He nods:* JUDITH *appears to the music of an adagio. In her right hand is a sword, in her left the head of* HOLOFERNES. *Dumb show as above.*

DUKE. I thank you for the pleasure you have given me. You are my guest from to-day; and for as long as you stay in my Court you are my close companion. Follow me, for I wish to introduce you to my wife as the most renowned magician.

FAUST. I obey, your Highness. [*They go out.*

SCENE IV

CASPER *and* MEPHISTOPHELES *enter.*

MEPHISTOPHELES [*dragging in Casper by the neck*]. Just you come here, you scoundrel! Why have you betrayed our master?

CASPER. Oh! oh! I haven't betrayed him. Let me alone, Stoffelfuss, golden Stoffelfuss—won't you let me go?

MEPHISTOPHELES. Why did you tell your master's name to the Duke?

CASPER. I didn't tell him a word; I just showed it to him pantomimically, and he understood at once who our master was and what was his name. But, dear Stoffelfuss, just let me go this time! Don't break my neck; I won't do it again as long as I live.

MEPHISTOPHELES. Well, it shall be allowed to pass this time. But as a punishment you will remain here in Parma alone. Master has dismissed you from his service. You'll see now how you'll perish! [*Goes out.*

CASPER [*alone*]. Stoffelfuss! He! Stoffelfuss! Golden Stoffel! Don't leave me alone! Stoffelchen! He's gone, by my soul, and left me here! First give me at least my wages! You owe me two months' wages. It's all in vain! The devil has taken it! Ah! You poor Casper, how will you get on now, without a place, without a master? And the Duke is sending round four men with big sticks to arrest me for my magic. [*Weeps.*] Hu! hu! ha! ha! If my grandmother knew what's happening to me I think the good woman would weep her eyes out of her head! Hu! hu! ha! ha!

Scene V

Auerhahn *enters.*

Auerhahn [*descending from the sky*]. Casper!

Casper. I thought some one called me by name?

Auerhahn. Casper, why are you in such grief?

Casper. And shouldn't I be grieved? My master has chucked me out of my place. Here I am in a foreign land where I don't know one street from another.

Auerhahn. You are truly in a difficult position, for here there are many bandits, who knock men dead for two halfpennies.

Casper. Two halfpennies? And I have just threepence in my pocket. Will they knock me dead thrice?

Auerhahn. Yes, they will knock you dead thrice!

Casper. Oh! oh! Poor Casper, all's over with you. Hu! hu! hu!

Auerhahn. Listen, Casper! I really pity you!

Casper. Well, here is one, at any rate, in the world who is affected by my position!

Auerhahn. Do you know? In Wittenberg the night-watchman has died, and if you promise me your soul, guaranteeing that I can carry it off after twelve years, I shall bring you to Wittenberg and put you in the night-watchman's place there.

Casper. No, nothing can come of this contract.

Auerhahn. Why not?

Casper. I haven't got a soul. My maker forgot to put one into me!

Auerhahn. Don't be so silly! Don't you consider yourself a man? Consequently you must have a soul.

Casper. Do you really believe I have a soul?

Auerhahn. Of course!

Casper [*aside*]. I can cheat the silly devil in this! [*Aloud*] All right! I remember now—I have got a soul! I don't know how I could have forgotten. But what's your name?

Auerhahn. Auerhahn!

Casper. Kickelhahn? Well, my dear Kickelhahn, I promise you my soul after twelve years—and you'll get me the night-watchman's place?

Auerhahn. Yes, I'll get it for you.

Casper. And bring me at once to Wittenberg?

Auerhahn. Yes, we shall be at Wittenberg in a twinkling. Just hold on to me.

CASPER [*grips him*]. I'm holding on! [*Springs back and blows on his hands.*] Thunder and lightning! I have burned myself; my hands must be full of blisters.

AUERHAHN. Yes, I've got an ardent disposition!

CASPER. I've noticed that. Just cool it off a little!

AUERHAHN. Well, then, hold on to me once more!

CASPER. Well, once more I'm holding on.

AUERHAHN. Say, now: Capo cnallo!

CASPER. Capers and quails.

AUERHAHN. Capo cnallo!

CASPER. Capo cnallo! [*Flies off with him.*

The curtain falls

ACT IV

SCENE I

FAUST'S *room as in Act I. The chair stands at the table.*

FAUST [*entering*]. Greetings, home of my earliest joy! Remove from me my depression, my ill-will! Oh, why have I renounced my hope of salvation for such an ordinary existence? Here is the wound whereby I subscribed my heart's blood to him, sure mortgage to hell. Of course, I may be of good hope, for I can laugh at Satan, since four and twenty years must pass after our contract before I become his bondslave, and now only half of that time has gone by. Yes, I shall make use of his help only for a few years more, to make myself famous, and then I shall endeavour to get out of his clutches, and seek to gain back the salvation I cast away so lightly on my abandoned path. [*He seats himself.*] But I do not know why such an overpowering desire for sleep has suddenly come on me and forces me here to take a rest. Ah! Rest—rest that since my bond has left me quite, so that up to now I have only known it as a name! [*He falls asleep.*

SCENE II

The GENIUS *enters.*

GENIUS. Faust! Faust! Wake from your sinful sleep! What have you undertaken? Consider that the joys which you gain from this infernal bond are transitory, that you have destroyed thereby your hope of salvation and go to eternal damnation! Were you not born a man, and do you sacrifice yourself so wantonly to this hellish spirit? Oh, abandon the road which you have been travelling up to now! Return to virtue! You have no time to lose if you desire still to save your soul. You can break the bond, but only if you do it to-day. Oh, Faust! Follow the warning of your guardian genius, so that I may flutter round protecting you as of yore! [*Goes out.*

FAUST [*awaking*]. Ha! What was that? This is the second time I thought I saw my genius warning me to break my bond with Satan as soon as possible. Yes, yes, I will go back to the path of virtue, and consecrate myself to it, and through it seek to make myself worthy of the joys of heaven. Mephistopheles!

SCENE III

MEPHISTOPHELES *enters.*

MEPHISTOPHELES. What you want, Faust?

FAUST. You know that you are forced by our contract to answer all my questions. Tell me then: what would you do if you could obtain salvation?

MEPHISTOPHELES. I am not compelled to give you an answer to such a question. Yet hear and despair! If I could gain eternal salvation I should climb a ladder all the way to heaven, even if every rung were a sharp knife. And do you, a man, throw your being so wantonly away in order to enjoy the transitory pleasures of earth?

FAUST. Ha! I am not yet in your power! Get away from me for ever!

MEPHISTOPHELES [*aside*]. Well, I will go away, and seek to bind him to me again by some means or other. [*Goes out.*

FAUST [*alone*]. Miserable wretch! How deep am I sunk! Yet there is still time to repent and to regain salvation. Yes, I will follow the words of my guardian spirit and at once relieve my heart by an ardent prayer to God. [*Kneels down.*] All-compassionate, look down from Your throne upon me, a sinning man. Listen to my sighs; let my prayer ascend through the clouds; forgive me my past sins; take me again into Your grace, and lead me . . .

SCENE IV

MEPHISTOPHELES *enters with* HELEN.

MEPHISTOPHELES. Faust! Faust! Leave off praying! Here, I'm bringing you the lovely Helen, for whom the whole of Troy was destroyed!

FAUST. Get away, you infernal spirit! I am in your power no more!

MEPHISTOPHELES. Just look here, Faust! She shall be your own, Faust, if only you stop praying!

FAUST [*looks round*]. Ha! What a charming shape do I see!

HELEN. Gracious sir, your huntsman told me you had some commands for me.

FAUST [*aside*]. Am I no longer myself? Ha! Are these my eyes which are devouring her eagerly and ardently as the sunbeams do the earth? Oh! The flame of life has blazed up in me anew; I shall try no more to gain heaven, for the earth blooms for me in amorous luxury.

MEPHISTOPHELES. Look, Faust, what trouble I've taken to dissipate your ill-humour! Amuse yourself with her as you please; only banish all sad thoughts.

FAUST. I thank you, Mephistopheles, for your lovely present. Now, charming Helen, are you desirous of living with me?

HELEN. You are lord of my person, and I will not fail to carry out your commands.

FAUST. Accompany me, dear Helen; I shall show you my jewels to convince you how happy your life with me will be.

HELEN. I follow you gladly! [*Goes out with* FAUST.

MEPHISTOPHELES [*alone*]. Ha! ha! ha! He's ours now; he's got no power now to escape us! He had almost overreached me and escaped from my clutches, but a woman was the thing to put him again into our hands.

SCENE V

FAUST *runs in.*

FAUST. Oh, vanity! Ha! Damned false being! When I sought to embrace this charming form I found myself embracing a hellish fury! Oh, Faust! What have you done? Now I have provoked Heaven anew! Once more I have allowed myself to be beguiled by Satan! Ha! Cursed spirit, are you still here? Get away from my side for ever, for I shall never see you again!

MEPHISTOPHELES. Ha! ha! ha! Rage on—it hurts me not! For know that our contract is nearly at an end; without any other chance of escape you will be my property.

FAUST. The contract at an end! I your property! Yet hardly half the four and twenty years have passed since I sold my soul to you!

MEPHISTOPHELES. No, Faust, you have made a bad mistake; just count in the nights and you will see that our contract is at an end.

FAUST. Ha! Lying spirit, you have betrayed me! But rejoice not so soon! I yet feel I have the power to defy you!

MEPHISTOPHELES. I laugh at your threats; your blood is mine; the bond is ended, and soon we shall come to take you in triumph to our prince Pluto. [*Goes out.*

SCENE VI

FAUST [*alone*]. Ha! Will my life's course then be ended in a few hours? [*Kneels down.*] Oh, may my prayer ascend yet once more to the all-good God! There where the rosy flames of evening soar, there is—ha! Curses! The fiery gate of hell! Listen! Never—there must I go—*Ave*—the music of the celestial choir is broken! Oh, demon, why do you twist my words so that my prayer is turned into curses? No, no, I cannot pray! The fountain of eternal mercy is sealed from me. Even if the angels were to weep tears on my account it would never be opened for me again! I can hope for mercy no more.

MEPHISTOPHELES [*within*]. Fauste, prepara te!

FAUST. Ha! Now must I prepare for the last hour of my life: now must I receive punishment for my sinful life—there in the pit of hell! My heart will be fettered by Pluto's heavy chains, and the furies wait eagerly for my body in order to tear it to pieces. [*It strikes ten.*

SCENE VII

CASPER *enters as a night-watchman with his lantern.*

CASPER [*still without*]. Grethel! Light the lantern for me. I must start my duties as night-watchman to-day. The citizens have just given me the job, but the town council hasn't yet confirmed it; it desires first a plain proof of my worth. Well, I'll do it as well as I can. [*Enters singing.*]

Masters all, now list to me:
If your wives they plaguy be
Into bed them straightway cast;
All the quarrel will be past.
Ten has struck.
Dra, la, la, la, la, la! [*Dances.*
Ladies all, now list to me,
You must bear much—that I see.
Yet this is no new device—
Sometime you've got to break the ice!
It's been broken quite a lot.
Dra, la, la, la, la, la! [*Dances.*

FAUST. How dare you enter my room when I have forbidden you ever to come to my house again?

CASPER. Your Insolence, pray pardon me; I desired only to give a proof of my skill as a night-watchman.

FAUST. Very good. But stay no longer in my presence.

CASPER. And then I wish also to talk to your Insolence about the wages owing to me; for I'm pressed for money. I must buy some trade equipment in order to carry on my new duties.

FAUST. Go to my famulus and get him to pay you the money—and now begone to the street where you belong.

CASPER. Well, if you are not glad to see me, I'll go. I thought to make my affairs right if I brought my first serenade to your Insolence's house, but since it doesn't appeal to you I'll make off again immediately. [*Goes out humming.*

FAUST [*alone*]. Now at this moment I am being accused and tried by the Almighty Judge! Oh, terrible thought!

MEPHISTOPHELES [*without*]. Fauste, judicatus es!

FAUST [*springs up*]. Ha! It is done! I am judged—my sentence is passed; the Almighty has broken his staff over me! I am in Satan's power! Oh, cursed be the day when I was born! [*Seats himself. It strikes eleven.*

SCENE VIII

FAUST, CASPER, *and later all the* FURIES.

CASPER [*within*]. Grethel, give me the lantern. It's struck again; the clock can't be quite right in its head. But pour me out first a little oil on the wick so that I can see a bit better. So! [*Enters singing.*

All ye widowers, list to me:
If a new wife you wish to see,
Do not praise the first too much,
Else you'll not get another such.
Eleven has struck!
Dra, la, la la, la, la!
All ye widows, list to me:
Truly you live in misery,
For, alas! you have not got
From experience you know what.
Eleven twenty!
Dra, la, la, la, la!

Zounds! How have I got into this room again! Pray pardon, your Insolence, that I have come into your house once more; I truly don't know how I manage to lose my path always! In a way, however, I'm glad that I have the honour of meeting your Insolence again, for I have a right big request to make to you.

FAUST. Well, what does your request concern?

CASPER. I have heard that your Insolence is to make a journey into the Plutonic realm, and I would wish to beg you to bear many compliments from me to my grandmother. She sits on the left hand as you enter hell, number one, and mends slippers.

FAUST. Get out this very minute, or I will drive your impertinence off by force!

CASPER. Well, well, don't take it so ill! I can easily go by myself. [*Goes out hummimg.*

MEPHISTOPHELES [*within*]. Fauste, in eternum damnatus es!

FAUST [*springs up*]. Ha! Now the moment has come when I am expected in the pit of hell, where resin and brimstone burns for me, where Pluto's monsters wait for me. Soon I shall feel hell's torments in my body! The thunder rolls—the earth vomits fire! Oh, help! Oh, save me, might of Heaven! In vain! In vain is my cry for help—I must hence to a place where I have to suffer punishment for my sins. Ha! Come then, you hellish furies, rend, tear my body, and bring me to the place of my fate!

[*At the beginning of this speech thunder and lightning start. These grow fiercer. At the end the* FURIES *arrive and go off carrying* FAUST *into the sky. Slowly it grows quieter, then twelve strikes.*

Scene IX

Casper. *Later* Auerhahn.

Casper [*without*]. Devil take it! That clock strikes as if Satan were pulling at the rope! I must go my rounds again. Grethel, make a couple of pans of coffee, but don't put too many grounds in it. Now give me the lantern. So! I will be back soon. See and be quick about it! [*Enters singing.*]

All my lads, now list to me!
If a maiden you go to see,
Do it nicely, do't with poise,
See the house door makes no noise:
Twelve has struck!
Dra, la, la, la, la!
All my virgins, list to me!
Should one ask you a question free—
"Are you, my dear, a virgin yet?"—
Just you answer: "Yes, I regret."
Null null has struck!
Dra, la, la, la, la!

[*He dances and bumps with the lantern into* Auerhahn, *who has descended from the sky.*

Casper [*shrieks*]. Kibi! Who's that?

Auerhahn. Do you not know me, Casper?

Casper. No, I don't! Who are you, *mon cher ami*?

Auerhahn. I am Auerhahn.

Casper. Oh! Let me just throw a little light on your face! [*Holds up the lantern.*] Yes, you're right. You are Kickelhahn. What do you want?

Auerhahn. Casper, your time is up. You must go with me to hell!

Casper. To hell? I thought the chimney was your place. I do believe you're not right in the head! Have the twelve years passed then?

Auerhahn. Yes.

Casper. But I made my first round as night-watchman only to-day.

Auerhahn. That's nothing. The twelve years have passed, and you are now mine.

Casper. What do you say? As I can see, you've cheated me!

Auerhahn. Of course!

Casper. Well, you're cheated too, for I haven't got a soul! Ha! ha! ha!

Auerhahn. And even if you haven't got a soul you must come with me.

Casper. Listen, Kickelhahn, don't make me wild! Go your way or my lantern will make companionship with your head!

Auerhahn. Well, you know that since you are a night-watchman I can't get you. [*Ascends through the sky.*

CASPER [*alone*]. That's charming—even the devil will have nothing to do with night-watchmen! Well, I'll go back to my comrades, and we'll make right merry with a can of schnaps and laugh at the silly devils. [*Goes out dancing.*

The curtain falls

FIG. 201. FIGURES FOR PUPPET-PLAYS

BIBLIOGRAPHY

WHILE this bibliography of books and articles is not intended to be exhaustive, I have considerably expanded the list of such works given in the original German edition of this book. The present bibliography provides, I think, a fuller list of available material on the theme than is elsewhere to be found.—*Translator.*

ABELS, H. R.: "Cinderella casts a Shadow" (*The School Arts Magazine* February 1931).

ACHARD, M.: "Guignol" (*Annales politiques et littéraires*, April 1926).

ACKLEY, E. F.: *Marionettes, Easy to make, Fun to use* (New York, 1929).

ALBER, —: *Les Théâtres d'ombres chinoises* (Paris, 1896).

ALBERT-BIROT, P.: *Matoum et Trévibar, ou Histoire édifiante et récréative du vrai et du faux poète* (Paris, 1919).

—— *Barbe-Bleue* (Paris, 1926).

ALFEROV, A.: Петрушка и его предки (Moscow, 1895).

ALTHERR, A.: *Marionetten* (Zürich, 1926).

AMIEUX, A.: *Cent ans après* (Lyons, 1904).

ANDERSON, M.: *The Heroes of the Puppet Stage* (1924).

B., O.: "Richard Teschner" (*Die graphischen Künste*, xli, 1918).

BAKSHY, A.: "The Lesson of the Puppet" (*The Theatre Arts Monthly*, July 1928).

BALDWIN, S.: "Dolls that come alive" (*The Woman's Home Companion*, December 1922).

BALLANTYNE, E.: "Sicilian Puppet-shows" (*The Theatre*, February 1893).

BALMER, H.: *Mein Gemüsetheater* (Bern, 1928).

BANNER, H. S.: "Java's Shadow-shows and the Kawi Epics" (*The London Mercury*, August 1927).

BARIL, G.: *Lafleur, Garçon apothicaire* (Amiens, 1901).

BARING, M.: "Punch and Judy" (*The London Mercury*, July 1922).

—— "Punch and Judy" (*The Living Age*, August 1922).

BEARD, L.: "A New Year's Punch-and-Judy Show" (*The Delineator*, January 1905).

BIBLIOGRAPHY

Beauplan, R. de: "Les Poupées animées de Ladislas Starévitch" (*La Petite Illustration*, March 1930).

Beckh, G. F.: *The Comedy of Marionettes, a Diary of Memories and Meditations* (n.d.).

Beissier, F.: *Théâtre de Guignol* (Paris, n.d.).

Belloc, H.: "Marionettes" (*The Outlook*, June 1923).

Berliner, R.: *Denkmäler der Krippenkunst* (Augsburg, 1926).

Bernard, L.: *Théâtre de Marionnettes* (texts, Paris, 1837).

Bernardelli, F.: "Per un teatro di marionette" (*Nuova Antologia*, April 1922).

Bertram, N. D.: "Das Puppentheater in der Sowjetunion" (*Das Puppentheater*, iii, pp. 161–168, 1929).

Bertrand, V.: *Les Silhouettes animées à la main* (Paris, 1892).

Bidon, H.: "Les petits Comédiens de bois" (*Journal des Débats*, January 1930).

Bielschowsky, A.: *Das Schwiegerlingsche Puppenspiel vom Doktor Faust* (Berlin, 1882).

Birrell, F.: "Puppets, The Tempest, and Mr Fagan" (*The Nation*, June 1923, with reply by J. B. Fagan).

Bistanclaque, —: *Guignol au Maroc* (Saint-Etienne, n.d.).

Bittner, K.: "Beiträge zur Geschichte des Volksschauspiels vom Doctor Faust" (*Prager deutsche Studien*, No. 27, 1922).

Blachetta, W.: *Blachetta-Spiele und andere* (Leipzig, 1930).

Blondeau, H., and Buteaux, V.: *Guignol s'en va-t-en guerre* (Paris, 1915).

Blümner, H.: "Fahrendes Volk im Altertum" (*Sitzungsberichte der k. bayerischen Akademie der Wissenschaften*, Munich; phil.-hist. Klasse, vi, 1918).

Bohatta, H.: "Das javanische Drama" (*Miteilungen der anthropol. Gesellschaft zu Wien*, 1905).

Bonaventure Batant: *Le Bottier de Saint-Georges* (Lyons, 1898).

Bonnaud, D.: *Pierrot pornographe* (Paris, 1902).

—— *Le Sacre de Clemenceau Ier* (Paris, 1907).

—— *Ulysse à Montmartre* (Paris, 1910).

—— *Venise, ou Lagune de miel* (Paris, 1913).

Bottcher, A.: "Vom Ausdruck des Kindes beim Handpuppenspiel" (*Blätter für Laien-und Jugendspieler*, i, 1925, 3).

Bouchor, M.: *Les Mystères d'Eleusis* (Paris, n.d.).

—— *Tobie* (Paris, n.d.).

—— *Noël, ou le Mystère de la Nativité* (Paris, n.d.).

—— *La Dévotion à Saint-André* (Paris, 1892).

—— *La Légende de Sainte-Cécile* (Paris, 1892).

—— *Le Songe de Khéyam* (Paris, 1892).

Boulton, W. B.: *The Amusements of Old London* (1901).

Bowie, A. G.: "The Story of Punch and Judy" (*The Theatre*, January 1884).

Brehm, W.: *Das Spiel mit der Handpuppe. Anleitung zur Herstellung von Handpuppen und Handpuppenbühnen und zum Spielen* (Düsseldorf, 1931).

Bresles, J. de: *Au Grand R. . Io . . Dé* (Dijon, 1924).

—— *Te rèves, eh Lyonnais!* (Dijon, 1925).

PUPPETS

BROOK, G. S.: "Memoirs of Marionettes" (*The Century Magazine*, March 1926).

BROWN, F. K.: "The Merrie Play of Punch and Judy" (*The Playground*, July 1921).

BRUINER, J. W.: *Faust vor Goethe* (Berlin, 1894).

BRUMME, M. A.: *Das kleine Theater* (Esslingen, 1926).

BRUX, G.: "Die theatergeschichtliche Bedeutung des Marionettenspiels" (*Jugendpflege*, iii, 1925, 1).

BUFANO, R.: "Puppet Anatomy" (*The Theatre Arts Monthly*, July 1928).

—— *Pinocchio for the Stage* (1929).

BUGNARD, C.: *L'École des ménagères* (Lyons, 1925).

BULLETT, G.: "Marionettes in Munich" (*The Saturday Review*, December 1929).

BULLY, M.: "The Return of the Marionettes" (*Current Opinion*, liv, March 1913).

CADILHAC, P.-É.: "Guignol à Paris" (*L'Illustration*, July 1930).

CALHOUN, L.: "Another Venture in Puppets" (*Drama*, October 1920).

CALTHROP, A.: "An Evening with Marionettes" (*The Theatre*, May 1884).

CALTHROP, D. C.: *Punch and Judy* (1926).

CALVI, E.: "Marionettes of Rome" (*The Bellman*, January 1917).

CANARD, G.: *Les Classiques du Gourguillon* (texts and introduction, Lyons, n.d.).

CANARD, G., DUROQUET, A., and COQUARD, G.: *Mémoires de l'Académie de Gourguillon* (Lyons, n.d.).

CANFIELD, M. C.: "Reflections on Tony Sarg's Marionettes" (*Vanity Fair*, April 1923).

CAVAZZA, E.: "At the Opra di li Pupi" (*The Atlantic Monthly*, June 1894).

CHANAY, A.: *L'Homme qui boit* (Lyons, 1924).

CHANCEL, J.: *Le Coffre-fort de Polichinelle* (Paris, n.d.).

CHAPUIS, A., and GÉLIS, E.: *Le Monde des automates* (Paris, 1928).

CHASE, E. F.: *Ballads in Black* (Boston, 1892).

CHESSÉ, R.: "Who will come to a Marionette Congress?" (*The Theatre Arts Monthly*, April 1931).

CHESTERTON, G. K.: *Tremendous Trifles* (1909).

CHÉZEL, F.: *Pierrot-Barnum* (Paris, 1902).

CHILD, T.: "A Christmas Mystery in the Fifteenth Century". (*Harper's Magazine*, December 1888).

CLAQUERET, I.: *Chantecoine, ou la Folie de Guignol* (Lyons, 1910).

CLAUDEL, P.: *L'Ours et la lune* (Paris, 1919).

COCHRANE, M. L.: "Japan's Doll Theatre, the Bunraku-za" (*Travel*, September 1923).

COLLIER, J. P.: *Punch and Judy* (1870).

COLOMBIER, H.: *Le Bandeau d' illusion* (Brussels, 1900).

CONANT, S. S.: "The Story of Punch and Judy" (*Harper's Monthly*, May 1871).

CONY, G.: *Manuel du Marionnettiste amateur* (Nice, n.d.).

COQUARD, G.: *Deux Artistes: Laurent Josserand, Henri Delisle* (Lyons, n.d.).

CORRERA, L.: "Il presepe a Napoli" (*L'Arte*, 1899).

COURTELINE, G.: *Marionetten* (Vienna, 1902).

CRAIG, E. GORDON: "School: an Interlude for Marionettes" (*The English Review*, January 1918).
—— "History" (*i.e.*, of the marionette stage) (*The Marionnette*, i, 1918, and ensuing numbers).
—— *Puppets and Poets* (1921).
—— "Marionettes and the English Press" (*The Mask*, 1929, p. 60).
CROZIÈRE, A.: *Le vrai Théâtre Guignol* (Paris, n.d.).
CUDDY, M. V.: "A Third-grade Project: a Puppet-show" (*Primary Education*, October 1927).
DARTHENAY, L.: *Le Guignol des salons* (Paris, 1888).
—— *Le Théâtre des petits* (Paris, 1890).
DAVID, É.: *El Bataille d' Querrin* (Amiens, 1891).
—— *Étude picarde sur Lafleur* (Amiens, 1896).
—— *Lafleur, ou le Valet picard* (Amiens, 1901).
—— *Lafleur en service* (Amiens, 1901).
—— *El Naissanche ed l'einfant Jésus* (Abbeville, 1905).
—— *Chés Histoires d'Lafleur* (Amiens, 1906).
—— *Les Théâtres populaires à Amiens* (Amiens, 1906).
—— *Vieilles réderies* (Amiens, 1920).
—— *Ch'viux Lafleur* (Amiens, 1926).
DAVIS, F. C.: "Story-telling by Means of Puppets" (*The Playground*, September 1926).
DELAUNAY, E.: *Guignol du grand cercle* (Aix-les-Bains, 1912).
DELVAU, A.: *Le Théâtre érotique français sous le Bas-Empire* (Paris, 1864).
DESNOYERS, F.: *Le Théâtre de Polichinelle* (Paris, 1861).
DESVERNAY, F.: *Laurent Mourguet et Guignol* (Lyons, 1912).
DEW, L. E.: "Amusing Children" (*Harper's Bazaar*, December 1910).
DILLEY, P.: "Burattini: Marionettes that are not Mechanical" (*Drama*, October–December 1923).
DOERING, O.: "Poccis Beziehungen zum Marionettentheater" (*Magdeburgische Zeitung*, 1903).
DONNAY, M.: *Phryné* (Paris, 1891).
—— *Ailleurs* (Paris, 1891).
—— *Autour du Chat noir* (Paris, 1926).
DOYEN, E.: *Les Marionnettes amoureuses* (Paris, n.d.).
DRESBACH, W.: "Designing a Simple Puppet-show" (*The School Arts Magazine*, January 1927).
DUCRET, É.: *Le Théâtre de Guignol* (Lyons, 1914).
DULBERG, F.: "Bühnensilhouetten. Schwabinger Schattenspiele" (*Zeitschrift für bild. Kunst*, 1908).
DUPLATEAU, M. (AUGUSTE BLETON): *Véridique Histoire de l'Académie de Gourguillon* (Lyons, 1918).
DURANTY, D.: *Théâtre des marionnettes du Jardin des Tuileries* (Paris, 1863).
DUROCHER, L.: *La Marche au soleil* (Paris, 1899).
DUVE, H.: "Die Wiedererweckung des Kasperletheaters" (*Westermanns Monatshefte*, February 1928).
EHRHARDT, G.: *Das Puppenspiel vom Doktor Faust* (Dresden, 1905).

Eisler, M.: "Richard Teschner" (*Dekorat. Kunst*, xxiv, 1921).

Endel, P.: *Ombres chinoises de mon père* (texts, Paris, 1885).

Engel, K., and Tille, A.: *Deutsche Puppenkomödien* (Oldenburg, 1879).

F., H. K.: "Das javanische Wayang-Schattenspiel" (*Neue Mannheimer Zeitung*, January 1925).

Fanciulli, G.: *Il teatro di Takiù* (texts, Milan, n.d.).

Ferny, J.: *Le Secret du manifestant* (Paris, 1893).

Ferrigni, P. C.: *La storia dei burattini* (Florence, 1902).

Feuillet, O.: *The Story of Mr Punch* (1929).

Finckh-Haelssig, M.: *Puppenschneiderei* (Ravensburg, 1928).

Flanagan, H.: "Puppets in Prague" (*The Theatre Arts Monthly*, April, May 1927).

Flögel, K. F.: *Geschichte des Groteskkomischen* (ed. M. Bauer, Munich, 1914).

Foa, É.: *Mémoires d'un Polichinelle* (Brussels, 1840).

Fournier, É.: *Histoire des jouets et des jeux d'enfants* (Paris, 1889).

Fragerolle, G.: *Le Rêve de Joël* (Paris, n.d.).

—— *Jeanne d'Arc* (Paris, n.d.).

—— *L'Aigle* (Paris, n.d.).

—— *L'Enfant prodigue* (Paris, 1894).

—— *Le Sphinx* (Paris, 1896).

—— *Clairs de lune* (Paris, 1896).

—— *Le Juif errant* (Paris, 1898).

Franck, P.: *Puppenspiele* (Berlin, 1931).

Frost, S. A.: *The Book of Tableaux and Shadow Pantomimes* (New York, 1869).

Fulda, Fr. W.: *Schattenspiele. Erfahrungen und Anregungen* (Rudolstadt, 1923).

Gabler, E. T.: "Marionetten" (*Kunst und Künstler*, 1926).

Gabriel, G. W.: "Opera on a Ten-foot Stage" (*Arts and Decoration*, December 1921).

Gaudefroy, L.: *Ech Mariage d' Lafleur* (Amiens, 1907).

Gentile, A. V.: *Teatrino per bambine e fanciuletti* (Milan, 1922).

—— *Burattini interessanti* (texts, Milan, 1925).

—— *Teatro per fanciulli e fanciulle* (Milan, 1925).

Ghelderode, M. de: *Le Mystère de la Passion de Notre-Seigneur Jésus-Christ* (Brussels, 1905).

Gibson, K.: "Shadow-plays" (*The School Arts Magazine*, March 1927).

Giradot, Mme: *Théâtre et Marionnettes pour les petits* (Paris, n.d.).

Gleason, A. H.: "The Last Stand of the Marionette" (*Collier's National Weekly*, October 1909).

Göbels, H., and Jünemann, J.: *Rulala, Rulala, Kasperle ist wieder da!* (Berlin, 1929).

Godart, J.: *Guignol et l'Esprit lyonnais* (Lyons, 1912).

—— *Guignol et la Guerre* (Lyons, 1919).

Göhler, C.: "Vom Kasperletheater" (*Kunstwart*, 1908).

Gonindard, J.: *Guignol locataire et la Chambre syndicale des propriétaires* (Lyons, 1894).

Goumard, J.: *Une Partie de billard du cercle des chefs d'atelier de la Rue de Crimée, à Lyon* (Lyons, 1914).

GRAFFIGNY, H. DE: *Le Théâtre à la maison* (Paris, n.d.).
—— *Construction du Théâtre Guignol* (Paris, n.d.).
GRAGGER, R.: "Deutsche Puppenspiele aus Ungarn" (*Archiv für das Studium der neueren Sprachen und Literaturen*, lxxx, 1925, 3–4).
GRÄSSE, R.: "Zur Geschichte des Puppenspiels und der Automaten" (in Romberg, *Geschichte des Wissenschaften im 19. Jahrh.*, Leipzig, 1856).
GREGORI, F.: "Marionetten- und Schattentheater" (*Kunstwart*, xx, pp. 361–362).
GRÖBER, K.: *Children's Toys of Bygone Days* (1928).
GRONEMANN, J.: "Das Meisseln der ledernen Wajang-Puppen der Javaner in der Vorstenlanden" (*Internat. Archiv für Ethnogr.*, xxi, 1913).
GROS, DOCTEUR (JOANNY BACHUT): *Pourquoi aimons-nous Guignol?* (Lyons, 1909).
GRUBE, W.: "Chinesische Schattenspiele" (*Sitzungsberichte der k. bayerische Akademie der Wissenschaften*; phil.-hist. Klasse 28, 1915).
GRUNEBAUM, M. R. V.: "Schattentheater und Scherenschnitt" (*Jahrbuch der österr. Leo-Gesellschaft*, 1929, pp. 141–177).
GUBALKE, L.: "Marionettentheater" (*Vom Fels zum Meer*, 1905, No. 17).
GUGITZ, G.: *Der weiland Kasperl* (Vienna, 1920).
GUIGNOLET, —: *Le Théâtre des ombres chinoises* (texts, Paris, n.d.).
HAEFKER, H.: "Vom Kasperletheater. Ein Stück Kulturgeschichte" (*Der Thürmer*, viii, 1905–06).
HAGEMANN, C.: *Die Spiele der Völker* (Berlin, 1921).
HAGER, G.: *Die Weihnachtskrippe* (Munich, 1902).
HALL, M. P.: "Java's Dancing Shadows" (*Overland Monthly*, July 1928).
HAMM, W.: *Das Puppenspiel vom Doctor Faust* (Leipzig, 1850).
HAMMOND, C. A.: "The Puppet-show" (*Hygeia*, June 1931).
HAMPE, T.: *Die fahrenden Leute in der deutschen Vergangenheit* (Leipzig, 1902).
HARAUCOURT, É.: *Héro et Léandre* (Paris, 1893).
HARTLAUB, G. F.: "Siamesische Schattenspiele" (*Die Woche*, xxvii, 1925, 28).
HAYES, J. J.: (edits puppet section of *Drama*).
HAZEU, G. A. J.: "Eine Wajang-Beber-Vorstellung in Jogjokarta" (*Internat. Archiv für Ethnogr.* xvi, 1904).
HEDDERWICK, T. C.: *The Old German Puppet-play of Doctor Faust* (1887).
HEINE, C.: *Das Schauspiel der deutschen Wanderbühne vor Gottsched* (Halle, 1889).
HEMPEL, O.: *Das Dresdner Kasperle* (Leipzig, 1931).
HILL, M.: "The Theatre of Once Upon a Time" (*Kindergarten and First Grade*, November 1921).
HINOT, C.: *Le Fils à Guignol* (Paris, n.d.).
HIRN, Y.: *Les Jeux d'enfants* (translation from the Swedish by T. Hammar, Paris, 1926).
HIRSCH, G.: "Puppet Performances in Germany" (*Harper's Weekly*, April 1916).
—— "A Master of Marionettes: Ernst Ehlert" (*Harper's Weekly*, April 1916).
HIRTH, F.: "Das Schattenspiel der Chinesen" (Budapest, *Keleti Szemle*, ii, 1901).
HOLROYD, M.: "The Marionette Theatre in Italy" (*The Nation*, September 1922).

HOLTHOF, L.: "Die Überreste des Goetheschen Puppentheaters und deren Geschichte" (*Freies deutsches Hochschrift*, 1882).

HORN, P.: "Das türkische Schattenspiel" (*Altgemeine Zeitung*, 112, 1900).

HOVER, O.: *Javanische Schattenspiele* (Leipzig, 1923).

HRBKOVA, S. B.: "Czechoslovak Puppet-shows" (*The Theatre Arts Monthly*, January 1923).

HUNDT, P.: *Deutsche Märchenspiele* (Oldenburg, 1922).

HUSSEY, D.: "Master Peter's Puppet-show" (*The Saturday Review*, November 1924).

IRVINE, J.: "Widow Polichinelle: our First Tragedienne addresses her Audience" (*Lippincott's Magazine*, February 1913).

IRWIN, E.: "Where the Players are Marionettes: a Little Italian Theatre in Mulberry Street" (*The Craftsman*, September 1907).

JACKSON, O. L.: "A Practical Puppet Theatre" (*The School Arts Magazine*, May 1924).

JACOB, G.: *Karagöz-Komödien* (Berlin, 1899).

—— *Das Schattentheater* (Berlin, 1901).

—— "Drei arabische Schattenspiele aus dem XIII. Jahrhundert" (Buda pest, *Keleti Szemle*, ii, 1901).

—— "Die wichtigsten älteren Nachrichten über das arabische Schattenspiel" (in Littman, E., *Arabische Schattenspiele*, Berlin, 1901).

—— *Bibliographie über das Schattentheater* (Erlangen, 1902).

—— *Geschichte des Schattentheaters* (Berlin, 1907).

—— "Chinesische Schattenschnitte" (*Cicerone*, xv, 1923, 22).

—— *Schattenschnitte aus Nordchina* (Berlin, 1923).

—— "Schatten- und Puppentheater im Orient" (*Der Bühnenvolksbund*, ii, 1926, 1).

—— *Dutangade, das ist wie der Affenprinz Angada als Gesandter auszog; ein altindisches Schattenspiel* (Leipzig, 1931).

JACQUIER, L.: *La Politique de Guignol, Gnafron, et Cie* (Lyons, 1876).

JACQUIN, J.: *La Prise de Pékin* (Paris, n.d.).

JARRY, A.: *Ubu sur la butte* (Paris, 1906).

JEANNE, P.: *Œdipe-Roi* (Paris, 1919).

—— *Bibliographie des Marionnettes* (Paris, 1926).

JENKINS, R. L.: "Industrial Art in Toyland" (*Arts and Decoration*, December 1922).

JEROME, L. B.: "Marionettes of Little Sicily" (*New England Magazine*, February 1910).

JHERING, H.: "Marionettentheater" (*Schaubuhne*, vi, 5).

JOSEPH, H. H.: *A Book of Marionettes* (1922, second edition 1931).

—— "The Figure Theatre of Richard Teschner" (*The Theatre Arts Monthly*, October 1923).

—— "Puppets of Brann and Puhonny" (*The Theatre Arts Monthly*, August 1924).

—— "Römische Marionetten" (*Das Puppentheater*, i, 1924, pp. 177–181).

—— *Ali Baba and Other Plays for Young People or Puppets* (New York, 1927).

—— "Pastoral Puppets" (*The Theatre Arts Monthly*, August 1929).

JUNGMANN, A. M.: "Marionettes Extraordinary" (*The Popular Science Monthly*, March 1918).

JUSSERAND, J. J.: "A Note on Pageants and Scaffolds Hye" (in *An English Miscellany presented to Dr Furnivall*, Oxford, 1901).

JUYNBOLL, H. H.: "Wajang Keletik oder Kerutjl" (*Internat. Archiv für Ethnogr.* xiii, 1900).

KAHLE, P. E.: *Zur Geschichte des arabischen Schattentheaters in Ägypten* (Leipzig, 1909).

—— "Islamische Schattenspielfiguren aus Ägypten" (*Der Islam*, i, 1910, pp. 264–299, and ii, 1911, pp. 143–195).

—— "Marktszene aus einem ägyptischen Schattenspiel" (*Zeitschrift für Assyriologie*, xxvii, 1912).

—— "Das islamische Schattentheater in Ägypten" (*Orientalisches Archiv*, iii, 1913, pp. 103–108).

—— "Das Krokodilspiel, ein ägyptisches Schattenspiel" (*Nachrichten der Göttinger Gesellschaft der Wissenschaften*, phil.-hist. Klasse, 1915).

—— "Das arabische Schattentheater in Ägypten" (*Blätter für Jugendspielscharen und Puppenspieler*, i, 1924, 1).

KAHN, G. (ed.): *Polichinelle . . . précedé d'une étude* (Paris, 1906).

KALB, D. B.: "Puppets" (*The School Arts Magazine*, November 1925, June 1927).

—— "Robinson Crusoe in Shadow-land" (*The School Arts Magazine*, May 1931).

KERN, F.: "Das ägyptische Schattentheater" (in Horowitz, J., *Spuren griechischer Mimen im Orient*, Berlin, 1905).

KINCAID, Z.: *Kabuki* (1925).

—— "Puppets in Japan" (*The Theatre Arts Monthly*, March 1929).

KING, G. G.: *Comedies and Legends for Marionettes* (New York, 1904).

KLEIST, H. VON: "A Marionette Theatre" (translation by D. M. McCollester, *The Theatre Arts Monthly*, July 1928).

KOLLMANN, A.: *Deutsche Puppenspiele* (Berlin, 1891).

KRALIK, R., and WINTER, S.: *Deutsche Puppenspiele* (Vienna, 1885).

KRAUS, A. V.: *Das böhmische Puppenspiel vom Doktor Faust* (Berlin, 1891).

KREYMBORG, A.: *Puppet-plays* (1923; preface by Gordon Craig).

—— "Writing for Puppets" (*The Theatre Arts Monthly*, October 1923).

KROTSCH, F.: *Künstler-Marionetten-Theater: Bildhauer Prof. Aichers, Salzburg* (Salzburg, 1926).

KÚNOS, M.: "Über türkische Schattenspiele" (*Ungarische Revue*, vii, 1887).

—— "Türkisches Puppentheater" (*Ethnologische Mitteilungen aus Ungarn*, ii, 1889).

LAGARDE, É.: *Ombres chinoises, guignols, et marionnettes* (Paris, 1900).

LAUBE, G., and LESEMANN, G.: *Schwarzweisskunst in der Hilfschule. Schattenspiele nach Märchen und verwandten Stoffen* (Halle, 1929).

LAWRENCE, W. J.: "The Immortal Mr Punch" (*The Living Age*, January 1921).

—— "Marionette Operas" (*The Musical Quarterly*, April 1924).

LEFEBVRE, É.: *Les Pièces de Théâtre guignol* (Lyons, 1912).

LEFFTZ, J.: "Puppenspeel im alten Strassburg" (*Elsassland*, 1929, pp. 233–237, 273–276).

LEGBAND, P.: "Die Renaissance der Marionette" (*Literarisches Echo*, ix, p. 248 *sqq.*).
LEHMANN, A: "Karagös" (*Das Puppentheater*, ii, 1925–26, pp. 46–47).
—— "Das deutsche Puppenspiel" (*Die vierte Wand*, 1927, 19).
LEHMANN, W.: "Ein türkische Schattenspiel" (*Die neue Orient*, iv, 1920, 2).
LEIBRECHT, G. P. J.: *Zeugnisse und Nachweise zur Geschichte des Puppenspiels in Deutschland* (Diss. Freiburg, 1919).
—— "Gesichtspunkte zu einer Geschichte des Puppenspiels" (*Das literarische Echo*, xxiii, 1920–21).
—— *Über Puppenspiele und ihre Pflege* (Innsbruck, 1921).
LEISEGANG, H. W.: "Marionettenspiel als künstler. Zeitausdruck" (*Das neue Reich*, xi, 1929).
LEMERCIER DE NEUVILLE: *Ombres chinoises* (texts, Paris, n.d.).
—— *Les Pupazzi au Chalet* (Vichy, 1865).
—— *I Pupazzi* (texts, Paris, 1866).
—— *Paris-Pantin* (texts, Paris, 1868).
—— *Le Théâtre des Pupazzi* (texts, Lyons, 1876).
—— *Les Pupazzi de l'enfance* (texts, Paris, 1881).
—— *Nouveau Théâtre des Pupazzi* (Paris, 1882).
—— *Histoire des Marionnettes modernes* (Paris, 1892).
—— *Les Pupazzi noirs: Ombres animées* (Paris, 1896).
—— *Les Pupazzi inédits* (Paris, 1903).
—— *Théâtre de marionnettes, à l'usage des enfants* (texts, Paris, 1904).
—— *Souvenirs d'un monteur de marionnettes* (Paris, 1911).
LEVIN, M.: "The Marionette Congress, 1930, Liége, Belgium" (*The Theatre Arts Monthly*, February 1931).
LEWIS, L. L.: "The Puppet-play as a Factor in Modern Education" (*Primary Education-Popular Educator*, June 1928).
LINK, O.: "Das Puppentheater in Rumänien" (*Das Puppentheater*, iii, 1929, pp. 173–176).
LITTMANN, E.: "Ein arabische Karagöz-Spiel" (*Zeitschrift der deutschen Morgenländischen Gesellschaft*, liv, 1900).
——*Arabische Schattenspiele* (Berlin, 1901).
—— "Das Malerspiel, ein Schattenspiel aus Aleppo, nach einer armenisch-türkischen Handschrift" (*Sitzungsberichte der Heidelberger Akademie der Wissenschaften;* phil.-hist. Klasse, viii, 1918).
LOEBER, J. A.: *Javaansche Shaduwbeelden* (Amsterdam, 1908).
LOKESCH, A. (ed.): *Alte Kasperlstücke* (texts, Berlin, 1931).
LORRAIN, J. DE: *La Barbe-Bleue* (Paris, n.d.).
LOVETT, L. S.: "Three Puppet-plays for a Rural School" (*The School Arts Magazine*, January 1931).
LOZOWICK, L.: "Exter's Marionettes" (*The Theatre Arts Monthly*, July 1928).
LUSCHAN, F. VON: "Das türkische Schattenspiel" (*Internat. Archiv für Ethnogr.* ii, 1889).
LUX, J. A.: "Geschichte und Ästhetik des Puppenspiels" (*Kind und Kunst*, 1906).
—— "Poccis Kasperlkomödien und die Marionettentheater" (*Allgemeine Zeitung*, Munich, 1909, No. 36).

BIBLIOGRAPHY

MacCarthy, D.: "Marionettes and Waxwork" (*The New Statesman*, April 1923).

Macdowall, H. S.: "The Faust of the Marionettes" (*The Living Age*, February 1901).

Mackay, C. D.: "Children's Plays in Italy" (*Drama*, October 1927).

—— "Puppet Theatres in Schools" (*Primary Education-Popular Educator*, September 1928).

Maeterlinck, M.: *Alladine et Palomides, Intérieur, et la Mort de Tintagiles, trois petits Drames pour marionnettes* (Brussels, 1894).

Magnin, C.: *Histoire des Marionnettes en Europe* (Paris, 1862).

Mahlmann, S. A.: *Marionettentheater* (Leipzig, 1806).

Maindron, É.: *Marionnettes et Guignols* (Paris, 1900).

Majut, R.: *Lebensbühne und Marionette* (Berlin, 1931).

Malamanni, V.: "Il teatro drammatico, le marionette e i burattini a Venezia nel secolo xviii" (*Nuova Antologia*, lxvii and lxviii).

Marzials, A: "Puppets as Pedagogues" (*The World's Children*, 1930).

Matthews, B.: "The Lamentable Tragedy of Punch and Judy" (*The Bookman*, December 1913).

—— "The Forerunner of the Movies" (*The Century Magazine*, April 1914).

—— "Puppet-shows, Old and New" (*The Bookman*, December 1914).

May, R.: "Aufbau eines Handpuppenspiels aus dem Stegreife" (*Volkspielkunst*, 1929).

Mayer, F. A.: "Beitrag zur Kenntnis des Puppentheaters" (*Euphorion*, vii, 1900).

Mayhew, H.: *London Labour and the London Poor* (1861).

Maz, G.: *Le Sarsifi Pétafiné* (Lyons, 1886).

McCabe, L. R.: "The Marionette Revival" (*The Theatre Magazine*, November 1920).

McCain, R.: "A Movable Playhouse" (*The Industrial Arts Magazine*, September 1919).

McCloud, N. C.: "Doll Play in a Doll Setting" (*The Mentor*, January 1928).

McIsaac, F. J.: "Tony Sarg" (*Drama*, December 1921).

—— "The Fun and Craft of the Puppet-show" (*The World Review*, March 1928).

McIsaac, F. J., and Stoddard, A.: *Marionettes and how to make Them* (1927).

McMillan, M. L.: "The Old, Old Story in Shadow-pictures" (*The Woman's Home Companion*, December 1925).

McPharlin, P.: *A Repertory of Marionette-plays* (New York, 1929).

—— "Anton Aicher's Marionette Theatre in Salzburg" (*Drama*, April 1929).

—— *Puppetry: A Year Book of Marionettes* (Detroit, 1930).

McQuinn, R.: "The Children's Theatre" (*The Delineator*, June 1919).

Menghini, M.: "Il teatro dei burattini; tradizioni cavalleresche romane" (*La cultura moderna*, 1911, pp. 39–42, 110–112).

Mentzel, E.: *Das Puppenspiel vom Erzzauberer Doktor Johann Faust* (Frankfort-on-the-Main, 1900).

Méthivet, —, and de Bussy, —: *Guignol musical* (Paris, n.d.).

Métivet, L.: *La Belle au bois dormant* (Paris, 1902).

—— *Aladin* (Paris, 1904).

MICHEL, W.: "Marionetten" (*Dekorative Kunst*, 1919).
MICK, H. L.: "Producing the Puppet-play" (*The Theatre Arts Monthly*, April 1921).
—— "Puppets, Here, There, and Elsewhere" (*Drama*, December 1922).
—— "The Face of a Puppet" (*Drama*, January 1923).
—— "How a Puppet gets his Head" (*Drama*, February 1923).
—— "Puppets from the Neck down" (*Drama*, April 1923).
—— "Dressing and stringing a Puppet" (*Drama*, May, June 1923).
MILLS, W. H., and DUNN, L. M.: *Marionettes, Masks, and Shadows* (1927).
MODERWELL, H. K.: "The Marionettes of Tony Sarg" (*Boston Transcript*, 1918).
MOENIUS, G.: "Münchener Marionettentheater" (*Münchener Allgemeine Rundschau*, xvi, 1929, 48).
MONNIER, M.: *Théâtre de Marionnettes* (texts, Geneva, 1871).
—— *Faust* (Geneva, 1871).
MONTOYA, G.: *Les Boers* (Paris, 1902).
MOULTON, R. H.: "Teaching Dolls to Act for Moving Pictures" (*The Illustrated World*, October 1917).
—— "Toyland in the Films" (*The Scientific American*, December 1917; *The Literary Digest*, February 1918).
MÜHLMANN, J.: "Alpenländische Weihnachtskrippen" (*Kunst und Kunsthandwerk*, xxiii, 1920).
MÜLLER, F. W. K.: "Siamesische Schattenspielfiguren im Kgl. Museum für Völkerkunde zu Berlin" (*Internat. Archiv für Ethnogr.* vii, 1894).
NASCIMBENI, G.: "Le commedie d'un burattinaio celebre" (*Il Marzocco*, xv, 9).
NAUMANN, H.: "Studien über das Puppenspiel" (*Zeitschrift für deutsche Bildung*, 1929, pp. 1–14).
NELSON, N., and HAYES, J. J.: "The Dancing Skeleton of a Marionette" (*Drama*, May 1927).
—— "Trick Marionettes" (*Drama*, October–December 1927, February–April 1928).
NERAD, H.: *Kasperl ist wieder da! Ein Wort für die Wiederbelebung des Handpuppentheaters* (Prague, 1922).
—— *Eine Studie bei Kaspar Hanswurst* (Prague, 1927).
NICOLAS, R.: "Le Théâtre d'ombres au Siam" (*The Journal of the Siamese Society*, 1927, pp. 37–52).
NICOLS, F. H.: "A Marionette Theatre in New York" (*The Century Magazine*, March 1902).
NIESSEN, C.: "Teatro dei Piccoli" (*Die vierte Wand*, 1927, 18).
—— "Marionetten-Theater" (*Velhagen und Klasings Monatshefte*, xlii, 1927–28).
—— *Das rheinische Puppenspiel* (Cologne, 1928).
NODIER, C.: *Contes de veillée* (Paris, 1856).
NOETH, O.: "Die Grazer Puppenspiele" (*Das Puppentheater*, i, 1923–24, pp. 129–135).
NOGUCHI, Y.: "The Japanese Puppet Theatre" (*Arts and Decoration*, October 1920).

Novelli, E.: *Il teatro dei burattini* (Milan, 1925; published anonymously).

Ohlendorf, H.: "Schattenspiele" (*Niedersochsen*, xxxiv, 1929).

Oncken, H.: *Kasper för Lütt un Grot. Anleitung zum Kasperspiel, Handpuppenspiel, Kartoffeltheater, Schattenspiel* (Oldenburg, 1931).

"Onofrio": *Théâtre lyonnais de Guignol* (texts and introduction, Lyons, 1865–70, latest edition 1910).

Painton, F. C.: "The Marionette as Correlator in the Public Schools" (*The School Arts Magazine*, December 1922).

Paltinieri, R.: *Il teatro dei piccoli* (texts, Milan, 1925).

Park, F.: "The Puppeteer's Library" (*The Theatre Arts Monthly*, July 1928).

Park, J. G.: "Puppets" (*The School Arts Magazine*, May 1924).

Parkhurst, W.: "Dead Actors for Live" (*Drama*, May 1919).

Patterson, A.: *Shadow Entertainments and how to work Them* (1895).

—— "The Puppets are coming to Town" (*The Theatre*, September 1917).

Paul, O.: *Neue Kasperspiele* (Leipzig, 1930).

—— (ed.): *Kasperstücke* (series of texts, Leipzig, n.d.).

Peixotto, E. C.: "Marionettes and Puppet-shows Past and Present" (*Scribner's Magazine*, March 1903).

Pennington, J.: "The Origin of Punch and Judy" (*The Mentor*, December 1924).

Peretz, V.: Кукольныи театръ на Руси (in Ежегодникъ Императорскихъ Театровъ, St Petersburg, 1895).

Petite, J. H. M.: *Guignols et Marionnettes* (Paris, 1911).

Petrai, G.: *Maschere e burattini* (Rome, 1885).

Petty, E.: "The Trail of the Long-nosed Princess" (*Drama*, April 1928).

Philippi, F.: *Schattenspiele* (Berlin, 1906).

Philippon, É.: *La Bernarda Buyandiri* (Lyons, 1885).

Picco, F.: "Lo scartafaccio di un burattinaio" (*Bolletino storico piacentino*, i, 1907).

Pierce, L. F.: "Punch and Judy Up-to-date" (*The World To-day*, March 1911).

—— "Successful Puppet-shows" (*The Theatre*, September 1916).

Pigeon, A.: *L'Amour dans les enfers* (Paris, n.d.).

Piper, M.: *Die Schaukunst der Japaner* (Berlin, 1927).

Pischel, R.: *Die Heimat des Puppenspiels* (Halle, 1900; translated by M. C. Tawney as *The Home of the Puppet-play*, 1902).

—— "Das altindische Schattenspiel" (*Sitzungsberichte der Berliner Akademie*, 1906).

Plimpton, E.: *Your Workshop* (New York, 1926).

Pocci, F.: *Lustiges Komödienbüchlein* (Munich, 1895).

—— *Heitere Lieder. Kasperliaden und Schattenspiele* (Munich, 1929).

Pollock, W. H.: "Marionettes" (*The Saturday Review*, August 1902).

Pougin, A.: *Dictionnaire du Théâtre* (Paris, 1885).

Poulsson, A. E.: "Shadow-plays" (*St Nicholas*, July 1907).

Powell, V. M.: "Říše Loutek, a Puppet Theatre in Prague" (*The Theatre Arts Monthly*, October 1930).

Poydenot, A.: *Polichinelle aux enfers* (Paris, 1899).

Prahlhaus, J.: *Kasperle-Speil* (Königsberg, 1927).

PROU, V.: "Les Théâtres d'automates en Grèce au Ie siècle avant l'ère chrétienne d'après les Αὐτοματοποιϊκά d'Héron d'Alexandrie" (*Mémoires presentés . . . à l'Academie*, I, ix, 1884).

PRÜFER, C.: "Das Schifferspiel" (*Beiträge zur Kenntnis des Orients*, ii, 1906, pp. 154–169).

—— *Ein ägyptisches Schattenspiel* (Erlangen, 1906).

PUHONNY, I.: "Marionettenkunst" (*Das Echo*, xl, 1923).

—— "The Physiognomy of the Marionette" (translation by H. H. Joseph, *The Theatre Arts Monthly*, July 1928).

QUEDENFELD, A.: "Das türkische Schattenspiel im Magrib" (*Ausland*, lxiii, 1900).

QUENNELL, PETER: "The Puppet Theatre" (in *A Superficial Journey through Tokyo and Peking*, London, 1932).

RABE, J. E.: *Kaspar Putschenelle. Historisches über die Handpuppen und althamburgische Kasperszenen* (Hamburg, 1924).

RACCA, C.: *Burattini e marionette* (Turin, 1925).

RACKY, J. (ed.): *Das Puppenspiel vom Doktor Faust* (Paderborn, 1927).

RAGUSA-MOLERTI, G.: "Una sacra rappresentazione in un teatro di marionette" (*Psiche*, xiii, 3).

RANSON, P.: *L'Abbé Prout* (texts, Paris, 1902).

RAPP, E.: *Die Marionetten in der deutschen Dichtung vom Sturm und Drang bis zur Romantik* (Leipzig, 1924).

REED, W. T.: "Puppetry" (*The Playground*, June 1930).

REHM, H. S.: *Das Buch der Marionetten* (Berlin, 1905).

REICH, H.: *Der Mimus* (Berlin, 1903).

REIGHARD, C.: *Plays for People and Puppets* (New York, 1928).

RESSEL, M., and LEONHARD, P. R.: *Seid ihr alle da? 10 lustige Stücke für das Kasperle-Theater* (Muhlhausen, 1931).

RHEDEN, K. VON: "Schattenspiele" (*Velhagen und Klasings Monatshefte*, 1908).

RHEIN, J.: "Mededeeling omtrent de chineesche Poppenkast" (*Internat. Archiv für Ethnogr.*, ii, 1889).

RICCI, C.: "I burattini in Bologna" (*La lettura*, iii, 11).

RICHMOND, E. T.: *Punch and Judy* (n.d.).

RIDGE, L.: "Kreymborg's Marionettes" (*The Dial*, January 1919).

RIESS, R.: "Die Marionettenbühne" (*Gegenwart*, May 1917).

RITTER, H.: *Karagös* (Hanover, 1924).

RIVIÈRE, H.: *La Tentation de Saint-Antoine* (Paris, 1887).

—— *La Marche à l'étoile* (Paris, 1890).

ROBERTS, C.: "Pulcinella" (*The Living Age*, April 1922).

ROHDEN, P. R.: *Das Puppenspiel* (Hamburg, 1923).

ROSE, A.: *The Boy Showman* (New York, 1926).

ROUSSEAU, V.: "A Puppet-play which lasts Two Months" (*Harper's Weekly*, October 1908).

ROUSSET, P.: *Un Divorce inutile* (Lyons, n.d.).

—— *Théâtre lyonnais de Guignol* (texts, Lyons, 1895).

ROZE, A.: "A Profile Puppet-show" (*The Scientific American*, May, June 1910).

RUSSELL, E.: "The Most Popular Play in the World" (*Outing Magazine*, January 1908).

RUTHENBURG, G. D.: "The Gooseberry Mandarin" (*The Theatre Arts Monthly*, July 1928).

SACHOIX, L., and VERRIÈRES, J. DES: *Chante-clair Guignol* (Lyons, 1912).

—— *Cyrano-Guignol* (Lyons, n.d.).

SAND, A.: "Les Marionnettes de Nohant" (*Annales politiques et littéraires*, October 1923).

SAND, M.: *Le Théâtre des Marionnettes* (texts, Paris, 1890).

SANDFORD, A.: "Books about Marionettes" (*Library Journal*, November 1929).

SARG, T.: *The Tony Sarg Marionette Book* (New York, 1921).

—— "Domesticating an Ancient Art" (*The Delineator*, April 1922).

—— "How to make and operate a Marionette Theatre" (*The Ladies' Home Journal*, December 1927).

—— "The Puppet-play in Education" (*Kindergarten and First Grade*, December 1924).

—— "The Revival of the Puppet-play in America" (*The Theatre Arts Monthly*, July 1928).

SAUNDERS, M. J.: "A Marionette-play in Four Acts" (*The School Arts Magazine*, January 1931).

SCHELL, S.: "Czech Puppets with a History" (*Shadowland*, January 1923).

SCHERRER, H.: "Statisches zur Geschichte des St Galler Marionettentheaters" (*Das Puppentheater*, i, 1923, 4).

—— "Wege und Ziele des St Galler Marionettentheaters" (*Das Puppentheater*, i, 1923, 4).

SCHEUERMANN, E.: *Handbuch der Kasperei. Vollständ. Lehrbuch des Handpuppenspiels* (Buchenbach, 1924).

—— *Neue Kasperstücke* (series of texts, Leipzig, n.d.).

SCHINK, J. F.: *Marionettentheater* (Munich, 1778).

SCHLEGEL, G.: *Chinesische Bräuche und Spiele in Europa* (Diss. Jena, 1869).

SCHMID-NOERR, F. A.: "Maske und Marionette" (*Pastor Bonus*, April 1929).

SCHMIDT, F. H.: *Moderne Marionettenspiele* (Leipzig, 1927).

—— *Allerlei Kasparstücke. Eine Bibliographie des Handpuppentheaters* (Leipzig, 1929).

SCHMIDT, O., and HOMANN, H.: *Handpuppenspiele* (Mühlhausen, 1930).

SCHMIDT, W.: "Heron von Alexandria" (*Neue Jahrbücher für das klassische Altertum*, ii, 1899).

SCHNEIDER, N. H.: *The Model Vaudeville Theatre* (1909).

SCHOTT, G.: *Die Puppenspiele des Grafen Pocci* (Frankfort-on-the-Main, 1911).

SCHULENBURG, W. VON: "Von der nationalen Mission des Puppentheaters" (*Das literarische Echo*, xix, 1917).

SCHÜRMANN-LINDNER, H.: "Das Puppenspiel in Paris" (*Das Puppentheater*, iii, 1929, 168–172).

—— "Vom Wesen des Puppenspiels" (*Heimblätter*, vi, 1929, pp. 272–277).

SCHWARZ, A.: *Neues Kasperl-Theater* (a series of plays, Leipzig, 1928).

SCHWITTAY, P.: "Wie führe ich Schattenspiele auf?" (*Optik und Schule*, 1929, pp. 17–25).

SÉGARD, C.: *Guignol apothicaire* (Paris, n.d.).

SEIF, T.: "Drei türkische Schattenspiele" (*Le Monde oriental*, xvii, 1923).

SELDES, G.: "Grock and Guignols" (*The New Republic*, April 1926).

(SÉRAPHIN): *Le Théâtre de Séraphin depuis son origine jusqu'à sa disparition* (Paris, 1872; published anonymously).

SERRURIER, L.: *De Wajang Poerwa, eene ethnologische Studie* (Leyden, 1896).

SEYBOLD, C. F.: "Zum arabischen Schattenspiel" (*Zeitschrift der deutschen Morgenländischen Gesellschaft*, lvi, 1902).

SHANKS, E.: "Puppetry and Life" (*The Outlook*, November 1923).

SHULTS, J. H.: "Teaching History by Puppets (*The Kindergarten Magazine*, September 1908).

SIBLEY, H.: "Marionettes, the Ever-popular Puppet-shows" (*Sunset*, November 1928).

SILVESTRE, A.: *Chemin de croix* (Paris, n.d.).

SIMOVICH, EFIMOVA: Запискiи Петрушечника (Moscow, 1925).

SMITH, W.: "Home Plays with Puppets" (*The Children's Royal*, December 1921).

SOMM, H.: *La Berline de l'émigré, ou jamais trop tard pour bien faire* (Paris, 1885).

SORBELLI, A.: "Angelo Cuccoli e le sue commedie" (*L'Archiginnasio*, iv).

STEMMLE, R. A. (ed.): *Das Handpuppentheater. Eine Reihe alter und neuer Komödien für die Handpuppenbühne* (a series of plays, Berlin, 1929).

STEVENSON, R. L.: *Memories and Portraits* (1887, with the famous essay "A Penny Plain and Twopence Coloured").

STODDART, A.: "The Renaissance of the Puppet-play" (*The Century Magazine*, June 1918).

STODDART, A., and SARG, T.: *A Book of Marionette-plays* (1930).

STRAUS, H.: "Puppet and Conductor" (*The Nation*, New York, February 1926).

STUDYNKA, F.: *Der rote Kasperl. Anleitung zur Kasperlstudien* (Vienna, 1929).

SURVILLE, G. DE: *Le Déménagement de Guignol* (Paris, n.d.).

SÜSSHEIM, K.: "Die moderne Gestalt des türkischen Schattenspiels (Quaragös)" (*Zeitschrift der deutschen Morgenländischen Gesellschaft*, lxiii, 1909).

SYMONS, A.: "An Apology for Puppets" (*The Saturday Review*, July 1897).

TALBOT, P. A.: "Some Magical Plays of Savages" (*The Strand Magazine*, June 1915).

TARDY, T.: *Profession libérale* (Dijon, 1926).

TAVERNIER, A., and ALEXANDRE, A.: *Le Guignol des Champs-Élysées* (preface by J. Claretie, Paris, n.d.).

TESAREK, A.: *Kasperl sucht den Weihnachtsmann* (Vienna, 1927).

THALASSO, A.: *Molière en Turquie: Étude sur le théâtre de Karagueuz* (Paris, 1888).

TICHENOR, G.: "Marionette Furioso: a Marionette-show in the House of Manteo" (*The Theatre Arts Monthly*, December 1929).

TIENNET, G.: *Le Rapide n° 6* (Lyons, n.d.).

TOLDO, P.: "Nella baracca dei burattini" (*Giornale storico della letteratura italiana*, li, 1908).

TRENARD, F.: *Quatre Pièces faciles à jouer* (Paris, n.d.).

TURNER, W. J.: "Marionette Opera" (*The New Statesman*, May 1923).

TUSSENBROEK, O. VAN: *De Toegepaste Kunsten in Nederland* (Rotterdam, 1905).

UNDERHILL, G.: "A New Field for Marionettes" (*Drama*, March 1924).

VERMOREL, J.: *Quelques petits théâtres lyonnais des XVIIIe et XIXe siècles* (Lyons, 1918).

VISAN, T. DE: *Le Guignol lyonnais* (preface by J. Clarétie, Paris, 1912).

WALLNER, E.: *Schattentheater* (Erfurt, 1895).

WALTERS, M. O.: "Puppet-shows for Primary Grades" (*Primary Education*, September 1925).

WALZ, J. A.: "Notes on the Puppet-play of Doctor Faust" (*Philological Quarterly*, July 1928).

WARSAGE, R. DE: *Histoire du célèbre Théâtre liégeois de marionnettes* (Brussels, 1905).

WEED, I.: "Puppet-plays for Children" (*The Century Magazine*, March 1916).

WEISMANTEL, L.: *Das Merkbuch der Puppenspiele* (Frankfort-on-the-Main, 1924).

—— *Das Schattenspielbuch* (Augsburg, 1929).

—— *Das Buch der Krippen* (Augsburg, 1930).

—— *Schattenspiele des weltliches und geistliches Jahres* (Augsburg, 1931).

WELLS, C. F.: "Puppet-shows" (*The Playground*, November 1929).

—— "Marionettes, Quaint Folk" (*The World Outlook*, October 1917).

WERNTZ, C. N.: "The Marionette Theatre of Japan" (*Our World*, April 1924).

WHANSLAW, H. W.: *Everybody's Theatre* (1923).

WHEELER, E. J.: "Starling Development of the Bi-dimensional Theatre" (*Current Literature*, May 1908).

WHIPPLE, L.: "Italy sends us Marionettes" (*The Survey*, April 1927).

WHITMIRE, L. G.: "Teaching School with Puppets" (*The World Review*, March 1928).

WILKINSON, W.: *The Peep Show* (1927).

—— *Vagabonds and Puppets* (1930).

—— *Puppets in Yorkshire* (1931).

WIMSATT, G.: "The Curious Puppet-shows of China" (*Travel*, December 1925).

WITTICH, E.: "Fahrende Puppenspieler" (*Schweizer. Archiv für Volkskunde*, 1929, pp. 54–61).

WOOD, E. H.: "Marionettes at Camp" (*The Playground*, March 1930).

WOOD, R. K.: "Puppets and Puppeteering" (*The Mentor*, April 1921).

WOODENSCONCE, PAPERNOSE: *The Wonderful Drama of Punch and Judy* (1919).

YAMBO, —: *Il teatro dei burattini* (texts, Milan, 1925).

YEATS, J. B.: *Plays for the Miniature Stage* (n.d.).

YOUNG, S. G.: "Guignol" (*Lippincott's Magazine*, August 1879).

ZACCO, T.: *Cenni storici sui fantocci, sulle marionette e su altri giocucci da fanciulli degli antichi* (Este, 1853).

ZEIGLER, F. J.: "Puppets, Ancient and Modern" (*Harper's Monthly*, December 1897).

ZELLNER, H.: "Vom Puppentheater zum Heimkino" (*Westermanns Monatshefte*, December 1929).

ZIMMERMANN, O.: *So baue ich mir ein Kasperltheater* (Leipzig, n.d.).

ZWIENER, B.: *Hallo-Hallo! Hier Kartoffeltheater* (Leipzig, 1928).

—— *Das neue Schattenspiel im Freien* (Leipzig, 1928).

—— *Die neue Schattenspiele daheim* (Leipzig, 1928).

PUPPETS

Miscellaneous

"Cassetta de' Burattini" (*The Penny Magazine*, March, April 1845).

"The History of Puppet-shows in England" (*Sharpe's London Journal*, July, December 1851).

"The Pedigree of Puppets" (*Household Words*, January 1852).

"Puppets of All Nations" (*Blackwood's Magazine*, April 1854).

"The Harlequinade" (*Chambers's Journal*, November 1856).

"Puppets, Religious and Aristocratic" (*Chambers's Journal*, December 1856).

"Popular Puppets" (*Chambers's Journal*, February 1857).

Neues Puppen-Theaters (Breslau, 1860).

Punch and Judy (1863, 1866, 1886, 1901).

Théâtre érotique de la rue de la Santé (texts, Brussels, 1864).

Mr Punch (1885).

"Puppet-shows" (*The Saturday Review*, March 1885).

Théâtre, Saynettes et Récits, par Gnafron fils . . . Neveu de Guignol (Lyons, 1886).

Guignol à la Comédie-Française. A propos d'une visite de Coquelin au Théâtre de Pierre Rousset (Lyons, 1887).

"A Greek Puppet-show" (*All the Year Round*, March 1894).

"A Puppet-show at the Paris Exhibition" (*The Scientific American*, November 1900).

"The Parisian Puppet Theatre" (*The Scientific American*, October 1902).

Papyrus et Martine, Punch et Judy, célèbre drame guignolesque anglais, pour la première fois adapté en France (Paris, 1903).

Les Parodies de Guignol: Répertoire de Pierre Rousset, Albert Chanay, Tony Tardy, Louis Josserand, Albert Avon (Lyons, 1911).

La Parodie de l'étranger (Lyons, 1913).

"The Return of the Marionets" (*Current Opinion*, March 1913).

Le Séraphin des enfants (texts, Epinal, 1914).

Que de Guignon! (Lyons, 1914).

"Punch and Judy" (*Current Opinion*, January 1914).

"The Most Immortal Character even seen on the Stage" (*Current Opinion*, January 1914).

"Puppet Warfare in France" (*The Literary Digest*, November 1915).

"Revival of the Puppets" (*Current Opinion*, July 1916).

"The Paradox of the Puppet: an Extinct Amusement born anew" (*Current Opinion*, July 1916).

The Marionnette (a journal edited by Gordon Craig, 1918).

"Are we forgetting Punch and Judy?" (*The Review of Reviews*, January 1918).

"How Puppets surpass our Human Actors: Tony Sarg's Marionettes" (*Current Opinion*, April 1918).

Guignol fait la guerre (Paris, 1919).

"Dolls knocking at the Actors' Door" (*The Literary Digest*, May 1919).

Im Kasperltheater (texts, Leipzig, 1919–22).

"Movies in the Time of William Shakespeare" (*Current Opinion*, May 1920).

"New Animation of the Inanimate Theatre" (*Vogue*, August 1920).

"Drama on Strings: Tony Sarg's Marionettes in Rip van Winkle" (*The Outlook*, December 1920).

BIBLIOGRAPHY

Das Puppenbuch (Berlin, 1921).
"Alice in Puppet Land" (*The Independent*, February 1921).
"Play-writing for the Puppet Theatre" (*Current Opinion*, May 1921).
"Resurrecting Chinese Movies a Thousand Years Old" (*Current Opinion*, July 1921).
La Malle (Paris, 1922).
"How Tony Sarg performs 'Miracles' with Marionettes" (*Current Opinion*, March 1922).
"Lilian Owen's Portrait Puppets" (*Drama*, October 1922).
"Guignol" (*The Nation*, April 1923).
"Portrait Puppets" (*Current Opinion*, April 1923).
Blätter für Jugendspielscharen und Puppenspieler (founded 1924).
"Rubber Actors lend Realism to Movies" (*The Popular Mechanics Magazine*, May 1924).
L'Entêtation amatée (Paris, 1925).
Nouveau Recueil de pièces de Guignol (Lyons, 1925).
"The Vogue for Puppet-plays" (*The Popular Educator*, January 1925).
"Behind the Scenes in a Puppet-show" (*The Popular Mechanics Magazine*, June 1925).
Vertingo (Paris, 1927).
Das Puppentheater-Modellierbuch (Berlin, 1927).
Vom Puppen- und Laienspiel (ed. W. Biel, Berlin, 1927).
"Our Puppet-show" (*Primary Education—Popular Educator*, January 1927).
"Les petits Comédiens de bois sur la scène du vieux Colombier" (*L'Illustration*, December 1927).
Wir Rüpelspieler (a series of plays, Berlin, 1927–28).
Eduard Blochs Kasperl-Theater (a series of plays, Berlin, 1927–30).
Das Kaspertheater des Leipziger Dürerbundes (a series of plays, Leipzig, 1927–30).
"Richard Teschner's Figure Theatre" (*The Theatre Arts Monthly*, July 1928).
"Telling the Story with Puppets" (*The Survey*, July 1928).
Funsterwalder Handpuppenspiele (a series of plays, Mühlhausen, 1928–29).
"Puppenspieler Pfingstfest in Prag" (*Das Puppentheater*, iii, 1929, pp. 129–138).
Handpuppenspiele (Mühlhausen, 1929).
Höflings Kasperl-Theater (a series of plays, Berlin, 1929–30).
Hohnsteiner Puppenspiele (a series of plays, edited by M. Jacob, Leipzig, 1929–30).
Höflings Schattentheater (a series of plays, Munich, 1929-31).
Radirullala, Kaspar ist wieder da! (a series of plays, Leipzig, 1929–31).
Kasperl-Theater (a series of plays, Berlin, 1929–31).
Kleine Kasparspiele (a series of plays, Leipzig, 1930).
Ollmärksche Puppenspeele (a series of plays, edited by O. Schulz-Heising, Leipzig, 1930).
Der rote Kasper (a series of plays, Leipzig, 1930–31).
"Puppets—What are They?" (*The Literary Digest*, January 1931).
Das alte Kölner Hänneschen-Theater (Cologne, 1931) (includes contributions by C. Niessen, O. Nettscher, R. Just, M. Hehemann, O. Klein, F. F. Wallraf, and H. Lindner).

PUPPETS

"Das Marionettentheater Münchener Künstler" (*Deutsche Kunst und Dekoration*, xix, pp. 89–93).

Das Puppentheater (organ of the puppet-theatre section of the society zur Förderung der deutschen Theaterkultur).

Der Puppenspieler (organ of the Deutscher Bund für Puppenspiel).

Le Séraphin de l'enfance (texts, Nancy, n.d.).

Loutkar (organ of the Union Internationale des Marionnettes, Prague. This society has also published a number of pamphlets relating to puppets and puppet-showmen).

Petit Répertoire de Guignol (Nice, n.d.).

Punch and Puppets ("The British Standard Hand Books," No. 38, n.d.).

Illustrations of various puppets in *The Theatre Arts Monthly*, June 1925, September 1925, November 1925, December 1925, June 1926.

INDEX

INDEX

INDEX

INDEX

A CATALOGUE OF SELECTED DOVER BOOKS IN ALL FIELDS OF INTEREST

A CATALOGUE OF SELECTED DOVER BOOKS IN ALL FIELDS OF INTEREST

AMERICA'S OLD MASTERS, James T. Flexner. Four men emerged unexpectedly from provincial 18th century America to leadership in European art: Benjamin West, J. S. Copley, C. R. Peale, Gilbert Stuart. Brilliant coverage of lives and contributions. Revised, 1967 edition. 69 plates. 365pp. of text.
21806-6 Paperbound $3.00

FIRST FLOWERS OF OUR WILDERNESS: AMERICAN PAINTING, THE COLONIAL PERIOD, James T. Flexner. Painters, and regional painting traditions from earliest Colonial times up to the emergence of Copley, West and Peale Sr., Foster, Gustavus Hesselius, Feke, John Smibert and many anonymous painters in the primitive manner. Engaging presentation, with 162 illustrations. xxii + 368pp.
22180-6 Paperbound $3.50

THE LIGHT OF DISTANT SKIES: AMERICAN PAINTING, 1760-1835, James T. Flexner. The great generation of early American painters goes to Europe to learn and to teach: West, Copley, Gilbert Stuart and others. Allston, Trumbull, Morse; also contemporary American painters—primitives, derivatives, academics—who remained in America. 102 illustrations. xiii + 306pp. 22179-2 Paperbound $3.00

A HISTORY OF THE RISE AND PROGRESS OF THE ARTS OF DESIGN IN THE UNITED STATES, William Dunlap. Much the richest mine of information on early American painters, sculptors, architects, engravers, miniaturists, etc. The only source of information for scores of artists, the major primary source for many others. Unabridged reprint of rare original 1834 edition, with new introduction by James T. Flexner, and 394 new illustrations. Edited by Rita Weiss. 6⅝ x 9⅝.
21695-0, 21696-9, 21697-7 Three volumes, Paperbound $13.50

EPOCHS OF CHINESE AND JAPANESE ART, Ernest F. Fenollosa. From primitive Chinese art to the 20th century, thorough history, explanation of every important art period and form, including Japanese woodcuts; main stress on China and Japan, but Tibet, Korea also included. Still unexcelled for its detailed, rich coverage of cultural background, aesthetic elements, diffusion studies, particularly of the historical period. 2nd, 1913 edition. 242 illustrations. lii + 439pp. of text.
20364-6, 20365-4 Two volumes, Paperbound $6.00

THE GENTLE ART OF MAKING ENEMIES, James A. M. Whistler. Greatest wit of his day deflates Oscar Wilde, Ruskin, Swinburne; strikes back at inane critics, exhibitions, art journalism; aesthetics of impressionist revolution in most striking form. Highly readable classic by great painter. Reproduction of edition designed by Whistler. Introduction by Alfred Werner. xxxvi + 334pp.
21875-9 Paperbound $2.50

Visual Illusions: Their Causes, Characteristics, and Applications, Matthew Luckiesh. Thorough description and discussion of optical illusion, geometric and perspective, particularly; size and shape distortions, illusions of color, of motion; natural illusions; use of illusion in art and magic, industry, etc. Most useful today with op art, also for classical art. Scores of effects illustrated. Introduction by William H. Ittleson. 100 illustrations. xxi + 252pp.
21530-X Paperbound $2.00

A Handbook of Anatomy for Art Students, Arthur Thomson. Thorough, virtually exhaustive coverage of skeletal structure, musculature, etc. Full text, supplemented by anatomical diagrams and drawings and by photographs of undraped figures. Unique in its comparison of male and female forms, pointing out differences of contour, texture, form. 211 figures, 40 drawings, 86 photographs. xx + 459pp. 5⅜ x 8⅜.
21163-0 Paperbound $3.50

150 Masterpieces of Drawing, Selected by Anthony Toney. Full page reproductions of drawings from the early 16th to the end of the 18th century, all beautifully reproduced: Rembrandt, Michelangelo, Dürer, Fragonard, Urs, Graf, Wouwerman, many others. First-rate browsing book, model book for artists. xviii + 150pp. 8⅜ x 11¼.
21032-4 Paperbound $2.50

The Later Work of Aubrey Beardsley, Aubrey Beardsley. Exotic, erotic, ironic masterpieces in full maturity: Comedy Ballet, Venus and Tannhauser, Pierrot, Lysistrata, Rape of the Lock, Savoy material, Ali Baba, Volpone, etc. This material revolutionized the art world, and is still powerful, fresh, brilliant. With *The Early Work,* all Beardsley's finest work. 174 plates, 2 in color. xiv + 176pp. 8⅛ x 11.
21817-1 Paperbound $3.00

Drawings of Rembrandt, Rembrandt van Rijn. Complete reproduction of fabulously rare edition by Lippmann and Hofstede de Groot, completely reedited, updated, improved by Prof. Seymour Slive, Fogg Museum. Portraits, Biblical sketches, landscapes, Oriental types, nudes, episodes from classical mythology—All Rembrandt's fertile genius. Also selection of drawings by his pupils and followers. "Stunning volumes," *Saturday Review.* 550 illustrations. lxxviii + 552pp. 9⅛ x 12¼.
21485-0, 21486-9 Two volumes, Paperbound $10.00

The Disasters of War, Francisco Goya. One of the masterpieces of Western civilization—83 etchings that record Goya's shattering, bitter reaction to the Napoleonic war that swept through Spain after the insurrection of 1808 and to war in general. Reprint of the first edition, with three additional plates from Boston's Museum of Fine Arts. All plates facsimile size. Introduction by Philip Hofer, Fogg Museum. v + 97pp. 9⅜ x 8¼.
21872-4 Paperbound $2.00

Graphic Works of Odilon Redon. Largest collection of Redon's graphic works ever assembled: 172 lithographs, 28 etchings and engravings, 9 drawings. These include some of his most famous works. All the plates from *Odilon Redon: oeuvre graphique complet,* plus additional plates. New introduction and caption translations by Alfred Werner. 209 illustrations. xxvii + 209pp. 9⅛ x 12¼.
21966-8 Paperbound $4.00

DESIGN BY ACCIDENT; A BOOK OF "ACCIDENTAL EFFECTS" FOR ARTISTS AND DESIGNERS, James F. O'Brien. Create your own unique, striking, imaginative effects by "controlled accident" interaction of materials: paints and lacquers, oil and water based paints, splatter, crackling materials, shatter, similar items. Everything you do will be different; first book on this limitless art, so useful to both fine artist and commercial artist. Full instructions. 192 plates showing "accidents," 8 in color. viii + 215pp. 8⅜ x 11¼. 21942-9 Paperbound $3.50

THE BOOK OF SIGNS, Rudolf Koch. Famed German type designer draws 493 beautiful symbols: religious, mystical, alchemical, imperial, property marks, runes, etc. Remarkable fusion of traditional and modern. Good for suggestions of timelessness, smartness, modernity. Text. vi + 104pp. 6⅛ x 9¼.
20162-7 Paperbound $1.25

HISTORY OF INDIAN AND INDONESIAN ART, Ananda K. Coomaraswamy. An unabridged republication of one of the finest books by a great scholar in Eastern art. Rich in descriptive material, history, social backgrounds; Sunga reliefs, Rajput paintings, Gupta temples, Burmese frescoes, textiles, jewelry, sculpture, etc. 400 photos. viii + 423pp. 6⅜ x 9¾. 21436-2 Paperbound $4.00

PRIMITIVE ART, Franz Boas. America's foremost anthropologist surveys textiles, ceramics, woodcarving, basketry, metalwork, etc.; patterns, technology, creation of symbols, style origins. All areas of world, but very full on Northwest Coast Indians. More than 350 illustrations of baskets, boxes, totem poles, weapons, etc. 378 pp.
20025-6 Paperbound $3.00

THE GENTLEMAN AND CABINET MAKER'S DIRECTOR, Thomas Chippendale. Full reprint (third edition, 1762) of most influential furniture book of all time, by master cabinetmaker. 200 plates, illustrating chairs, sofas, mirrors, tables, cabinets, plus 24 photographs of surviving pieces. Biographical introduction by N. Bienenstock. vi + 249pp. 9⅞ x 12¾. 21601-2 Paperbound $4.00

AMERICAN ANTIQUE FURNITURE, Edgar G. Miller, Jr. The basic coverage of all American furniture before 1840. Individual chapters cover type of furniture—clocks, tables, sideboards, etc.—chronologically, with inexhaustible wealth of data. More than 2100 photographs, all identified, commented on. Essential to all early American collectors. Introduction by H. E. Keyes. vi + 1106pp. 7⅞ x 10¾.
21599-7, 21600-4 Two volumes, Paperbound $11.00

PENNSYLVANIA DUTCH AMERICAN FOLK ART, Henry J. Kauffman. 279 photos, 28 drawings of tulipware, Fraktur script, painted tinware, toys, flowered furniture, quilts, samplers, hex signs, house interiors, etc. Full descriptive text. Excellent for tourist, rewarding for designer, collector. Map. 146pp. 7⅞ x 10¾.
21205-X Paperbound $2.50

EARLY NEW ENGLAND GRAVESTONE RUBBINGS, Edmund V. Gillon, Jr. 43 photographs, 226 carefully reproduced rubbings show heavily symbolic, sometimes macabre early gravestones, up to early 19th century. Remarkable early American primitive art, occasionally strikingly beautiful; always powerful. Text. xxvi + 207pp. 8⅜ x 11¼. 21380-3 Paperbound $3.50

A HISTORY OF COSTUME, Carl Köhler. Definitive history, based on surviving pieces of clothing primarily, and paintings, statues, etc. secondarily. Highly readable text, supplemented by 594 illustrations of costumes of the ancient Mediterranean peoples, Greece and Rome, the Teutonic prehistoric period; costumes of the Middle Ages, Renaissance, Baroque, 18th and 19th centuries. Clear, measured patterns are provided for many clothing articles. Approach is practical throughout. Enlarged by Emma von Sichart. 464pp. 21030-8 Paperbound $3.50

ORIENTAL RUGS, ANTIQUE AND MODERN, Walter A. Hawley. A complete and authoritative treatise on the Oriental rug—where they are made, by whom and how, designs and symbols, characteristics in detail of the six major groups, how to distinguish them and how to buy them. Detailed technical data is provided on periods, weaves, warps, wefts, textures, sides, ends and knots, although no technical background is required for an understanding. 11 color plates, 80 halftones, 4 maps. vi + 320pp. 6⅛ x 9⅛. 22366-3 Paperbound $5.00

TEN BOOKS ON ARCHITECTURE, Vitruvius. By any standards the most important book on architecture ever written. Early Roman discussion of aesthetics of building, construction methods, orders, sites, and every other aspect of architecture has inspired, instructed architecture for about 2,000 years. Stands behind Palladio, Michelangelo, Bramante, Wren, countless others. Definitive Morris H. Morgan translation. 68 illustrations. xii + 331pp. 20645-9 Paperbound $3.50

THE FOUR BOOKS OF ARCHITECTURE, Andrea Palladio. Translated into every major Western European language in the two centuries following its publication in 1570, this has been one of the most influential books in the history of architecture. Complete reprint of the 1738 Isaac Ware edition. New introduction by Adolf Placzek, Columbia Univ. 216 plates. xxii + 110pp. of text. 9½ x 12¾. 21308-0 Clothbound $10.00

STICKS AND STONES: A STUDY OF AMERICAN ARCHITECTURE AND CIVILIZATION, Lewis Mumford.One of the great classics of American cultural history. American architecture from the medieval-inspired earliest forms to the early 20th century; evolution of structure and style, and reciprocal influences on environment. 21 photographic illustrations. 238pp. 20202-X Paperbound $2.00

THE AMERICAN BUILDER'S COMPANION, Asher Benjamin. The most widely used early 19th century architectural style and source book, for colonial up into Greek Revival periods. Extensive development of geometry of carpentering, construction of sashes, frames, doors, stairs; plans and elevations of domestic and other buildings. Hundreds of thousands of houses were built according to this book, now invaluable to historians, architects, restorers, etc. 1827 edition. 59 plates. 114pp. 7⅞ x 10¾. 22236-5 Paperbound $3.50

DUTCH HOUSES IN THE HUDSON VALLEY BEFORE 1776, Helen Wilkinson Reynolds. The standard survey of the Dutch colonial house and outbuildings, with constructional features, decoration, and local history associated with individual homesteads. Introduction by Franklin D. Roosevelt. Map. 150 illustrations. 469pp. 6⅝ x 9¼. 21469-9 Paperbound $4.00

THE ARCHITECTURE OF COUNTRY HOUSES, Andrew J. Downing. Together with Vaux's *Villas and Cottages* this is the basic book for Hudson River Gothic architecture of the middle Victorian period. Full, sound discussions of general aspects of housing, architecture, style, decoration, furnishing, together with scores of detailed house plans, illustrations of specific buildings, accompanied by full text. Perhaps the most influential single American architectural book. 1850 edition. Introduction by J. Stewart Johnson. 321 figures, 34 architectural designs. xvi + 560pp.
22003-6 Paperbound $4.00

LOST EXAMPLES OF COLONIAL ARCHITECTURE, John Mead Howells. Full-page photographs of buildings that have disappeared or been so altered as to be denatured, including many designed by major early American architects. 245 plates. xvii + 248pp. 7⅞ x 10¾. 21143-6 Paperbound $3.50

DOMESTIC ARCHITECTURE OF THE AMERICAN COLONIES AND OF THE EARLY REPUBLIC, Fiske Kimball. Foremost architect and restorer of Williamsburg and Monticello covers nearly 200 homes between 1620-1825. Architectural details, construction, style features, special fixtures, floor plans, etc. Generally considered finest work in its area. 219 illustrations of houses, doorways, windows, capital mantels. xx + 314pp. 7⅞ x 10¾. 21743-4 Paperbound $4.00

EARLY AMERICAN ROOMS: 1650-1858, edited by Russell Hawes Kettell. Tour of 12 rooms, each representative of a different era in American history and each furnished, decorated, designed and occupied in the style of the era. 72 plans and elevations, 8-page color section, etc., show fabrics, wall papers, arrangements, etc. Full descriptive text. xvii + 200pp. of text. 8⅜ x 11¼.
21633-0 Paperbound $5.00

THE FITZWILLIAM VIRGINAL BOOK, edited by J. Fuller Maitland and W. B. Squire. Full modern printing of famous early 17th-century ms. volume of 300 works by Morley, Byrd, Bull, Gibbons, etc. For piano or other modern keyboard instrument; easy to read format. xxxvi + 938pp. 8⅜ x 11.
21068-5, 21069-3 Two volumes, Paperbound $10.00

KEYBOARD MUSIC, Johann Sebastian Bach. Bach Gesellschaft edition. A rich selection of Bach's masterpieces for the harpsichord: the six English Suites, six French Suites, the six Partitas (Clavierübung part I), the Goldberg Variations (Clavierübung part IV), the fifteen Two-Part Inventions and the fifteen Three-Part Sinfonias. Clearly reproduced on large sheets with ample margins; eminently playable. vi + 312pp. 8⅛ x 11. 22360-4 Paperbound $5.00

THE MUSIC OF BACH: AN INTRODUCTION, Charles Sanford Terry. A fine, nontechnical introduction to Bach's music, both instrumental and vocal. Covers organ music, chamber music, passion music, other types. Analyzes themes, developments, innovations. x + 114pp. 21075-8 Paperbound $1.25

BEETHOVEN AND HIS NINE SYMPHONIES, Sir George Grove. Noted British musicologist provides best history, analysis, commentary on symphonies. Very thorough, rigorously accurate; necessary to both advanced student and amateur music lover. 436 musical passages. vii + 407 pp. 20334-4 Paperbound $2.75

JOHANN SEBASTIAN BACH, Philipp Spitta. One of the great classics of musicology, this definitive analysis of Bach's music (and life) has never been surpassed. Lucid, nontechnical analyses of hundreds of pieces (30 pages devoted to St. Matthew Passion, 26 to B Minor Mass). Also includes major analysis of 18th-century music. 450 musical examples. 40-page musical supplement. Total of xx + 1799pp.
(EUK) 22278-0, 22279-9 Two volumes, Clothbound $17.50

MOZART AND HIS PIANO CONCERTOS, Cuthbert Girdlestone. The only full-length study of an important area of Mozart's creativity. Provides detailed analyses of all 23 concertos, traces inspirational sources. 417 musical examples. Second edition. 509pp. (USO) 21271-8 Paperbound $3.50

THE PERFECT WAGNERITE: A COMMENTARY ON THE NIBLUNG'S RING, George Bernard Shaw. Brilliant and still relevant criticism in remarkable essays on Wagner's Ring cycle, Shaw's ideas on political and social ideology behind the plots, role of Leitmotifs, vocal requisites, etc. Prefaces. xxi + 136pp.
21707-8 Paperbound $1.50

DON GIOVANNI, W. A. Mozart. Complete libretto, modern English translation; biographies of composer and librettist; accounts of early performances and critical reaction. Lavishly illustrated. All the material you need to understand and appreciate this great work. Dover Opera Guide and Libretto Series; translated and introduced by Ellen Bleiler. 92 illustrations. 209pp.
21134-7 Paperbound $2.00

HIGH FIDELITY SYSTEMS: A LAYMAN'S GUIDE, Roy F. Allison. All the basic information you need for setting up your own audio system: high fidelity and stereo record players, tape records, F.M. Connections, adjusting tone arm, cartridge, checking needle alignment, positioning speakers, phasing speakers, adjusting hums, trouble-shooting, maintenance, and similar topics. Enlarged 1965 edition. More than 50 charts, diagrams, photos. iv + 91pp. 21514-8 Paperbound $1.25

REPRODUCTION OF SOUND, Edgar Villchur. Thorough coverage for laymen of high fidelity systems, reproducing systems in general, needles, amplifiers, preamps, loudspeakers, feedback, explaining physical background. "A rare talent for making technicalities vividly comprehensible," R. Darrell, *High Fidelity.* 69 figures. iv + 92pp. 21515-6 Paperbound $1.25

HEAR ME TALKIN' TO YA: THE STORY OF JAZZ AS TOLD BY THE MEN WHO MADE IT, Nat Shapiro and Nat Hentoff. Louis Armstrong, Fats Waller, Jo Jones, Clarence Williams, Billy Holiday, Duke Ellington, Jelly Roll Morton and dozens of other jazz greats tell how it was in Chicago's South Side, New Orleans, depression Harlem and the modern West Coast as jazz was born and grew. xvi + 429pp.
21726-4 Paperbound $2.50

FABLES OF AESOP, translated by Sir Roger L'Estrange. A reproduction of the very rare 1931 Paris edition; a selection of the most interesting fables, together with 50 imaginative drawings by Alexander Calder. v + 128pp. 6½x9¼.
21780-9 Paperbound $1.50

AGAINST THE GRAIN (A REBOURS), Joris K. Huysmans. Filled with weird images, evidences of a bizarre imagination, exotic experiments with hallucinatory drugs, rich tastes and smells and the diversions of its sybarite hero Duc Jean des Esseintes, this classic novel pushed 19th-century literary decadence to its limits. Full unabridged edition. Do not confuse this with abridged editions generally sold. Introduction by Havelock Ellis. xlix + 206pp. 22190-3 Paperbound $2.00

VARIORUM SHAKESPEARE: HAMLET. Edited by Horace H. Furness; a landmark of American scholarship. Exhaustive footnotes and appendices treat all doubtful words and phrases, as well as suggested critical emendations throughout the play's history. First volume contains editor's own text, collated with all Quartos and Folios. Second volume contains full first Quarto, translations of Shakespeare's sources (Belleforest, and Saxo Grammaticus), Der Bestrafte Brudermord, and many essays on critical and historical points of interest by major authorities of past and present. Includes details of staging and costuming over the years. By far the best edition available for serious students of Shakespeare. Total of xx + 905pp.
21004-9, 21005-7, 2 volumes, Paperbound $7.00

A LIFE OF WILLIAM SHAKESPEARE, Sir Sidney Lee. This is the standard life of Shakespeare, summarizing everything known about Shakespeare and his plays. Incredibly rich in material, broad in coverage, clear and judicious, it has served thousands as the best introduction to Shakespeare. 1931 edition. 9 plates. xxix + 792pp. (USO) 21967-4 Paperbound $3.75

MASTERS OF THE DRAMA, John Gassner. Most comprehensive history of the drama in print, covering every tradition from Greeks to modern Europe and America, including India, Far East, etc. Covers more than 800 dramatists, 2000 plays, with biographical material, plot summaries, theatre history, criticism, etc. "Best of its kind in English," *New Republic*. 77 illustrations. xxii + 890pp.
20100-7 Clothbound $8.50

THE EVOLUTION OF THE ENGLISH LANGUAGE, George McKnight. The growth of English, from the 14th century to the present. Unusual, non-technical account presents basic information in very interesting form: sound shifts, change in grammar and syntax, vocabulary growth, similar topics. Abundantly illustrated with quotations. Formerly *Modern English in the Making*. xii + 590pp.
21932-1 Paperbound $3.50

AN ETYMOLOGICAL DICTIONARY OF MODERN ENGLISH, Ernest Weekley. Fullest, richest work of its sort, by foremost British lexicographer. Detailed word histories, including many colloquial and archaic words; extensive quotations. Do not confuse this with the Concise Etymological Dictionary, which is much abridged. Total of xxvii + 830pp. 6½ x 9¼.
21873-2, 21874-0 Two volumes, Paperbound $6.00

FLATLAND: A ROMANCE OF MANY DIMENSIONS, E. A. Abbott. Classic of science-fiction explores ramifications of life in a two-dimensional world, and what happens when a three-dimensional being intrudes. Amusing reading, but also useful as introduction to thought about hyperspace. Introduction by Banesh Hoffmann. 16 illustrations. xx + 103pp. 20001-9 Paperbound $1.00

LAST AND FIRST MEN AND STAR MAKER, TWO SCIENCE FICTION NOVELS, Olaf Stapledon. Greatest future histories in science fiction. In the first, human intelligence is the "hero," through strange paths of evolution, interplanetary invasions, incredible technologies, near extinctions and reemergences. Star Maker describes the quest of a band of star rovers for intelligence itself, through time and space: weird inhuman civilizations, crustacean minds, symbiotic worlds, etc. Complete, unabridged. v + 438pp. 21962-3 Paperbound $2.50

THREE PROPHETIC NOVELS, H. G. WELLS. Stages of a consistently planned future for mankind. *When the Sleeper Wakes,* and *A Story of the Days to Come,* anticipate *Brave New World* and *1984,* in the 21st Century; *The Time Machine,* only complete version in print, shows farther future and the end of mankind. All show Wells's greatest gifts as storyteller and novelist. Edited by E. F. Bleiler. x + 335pp. (USO) 20605-X Paperbound $2.50

THE DEVIL'S DICTIONARY, Ambrose Bierce. America's own Oscar Wilde—Ambrose Bierce—offers his barbed iconoclastic wisdom in over 1,000 definitions hailed by H. L. Mencken as "some of the most gorgeous witticisms in the English language." 145pp. 20487-1 Paperbound $1.25

MAX AND MORITZ, Wilhelm Busch. Great children's classic, father of comic strip, of two bad boys, Max and Moritz. Also Ker and Plunk (Plisch und Plumm), Cat and Mouse, Deceitful Henry, Ice-Peter, The Boy and the Pipe, and five other pieces. Original German, with English translation. Edited by H. Arthur Klein; translations by various hands and H. Arthur Klein. vi + 216pp.
20181-3 Paperbound $2.00

PIGS IS PIGS AND OTHER FAVORITES, Ellis Parker Butler. The title story is one of the best humor short stories, as Mike Flannery obfuscates biology and English. Also included, That Pup of Murchison's, The Great American Pie Company, and Perkins of Portland. 14 illustrations. v + 109pp. 21532-6 Paperbound $1.25

THE PETERKIN PAPERS, Lucretia P. Hale. It takes genius to be as stupidly mad as the Peterkins, as they decide to become wise, celebrate the "Fourth," keep a cow, and otherwise strain the resources of the Lady from Philadelphia. Basic book of American humor. 153 illustrations. 219pp. 20794-3 Paperbound $1.50

PERRAULT'S FAIRY TALES, translated by A. E. Johnson and S. R. Littlewood, with 34 full-page illustrations by Gustave Doré. All the original Perrault stories—Cinderella, Sleeping Beauty, Bluebeard, Little Red Riding Hood, Puss in Boots, Tom Thumb, etc.—with their witty verse morals and the magnificent illustrations of Doré. One of the five or six great books of European fairy tales. viii + 117pp. 8⅛ x 11. 22311-6 Paperbound $2.00

OLD HUNGARIAN FAIRY TALES, Baroness Orczy. Favorites translated and adapted by author of the *Scarlet Pimpernel.* Eight fairy tales include "The Suitors of Princess Fire-Fly," "The Twin Hunchbacks," "Mr. Cuttlefish's Love Story," and "The Enchanted Cat." This little volume of magic and adventure will captivate children as it has for generations. 90 drawings by Montagu Barstow. 96pp.
(USO) 22293-4 Paperbound $1.95

THE RED FAIRY BOOK, Andrew Lang. Lang's color fairy books have long been children's favorites. This volume includes Rapunzel, Jack and the Bean-stalk and 35 other stories, familiar and unfamiliar. 4 plates, 93 illustrations x + 367pp.
21673-X Paperbound $2.50

THE BLUE FAIRY BOOK, Andrew Lang. Lang's tales come from all countries and all times. Here are 37 tales from Grimm, the Arabian Nights, Greek Mythology, and other fascinating sources. 8 plates, 130 illustrations. xi + 390pp.
21437-0 Paperbound $2.50

HOUSEHOLD STORIES BY THE BROTHERS GRIMM. Classic English-language edition of the well-known tales — Rumpelstiltskin, Snow White, Hansel and Gretel, The Twelve Brothers, Faithful John, Rapunzel, Tom Thumb (52 stories in all). Translated into simple, straightforward English by Lucy Crane. Ornamented with headpieces, vignettes, elaborate decorative initials and a dozen full-page illustrations by Walter Crane. x + 269pp. 21080-4 Paperbound $2.50

THE MERRY ADVENTURES OF ROBIN HOOD, Howard Pyle. The finest modern versions of the traditional ballads and tales about the great English outlaw. Howard Pyle's complete prose version, with every word, every illustration of the first edition. Do not confuse this facsimile of the original (1883) with modern editions that change text or illustrations. 23 plates plus many page decorations. xxii + 296pp.
22043-5 Paperbound $2.50

THE STORY OF KING ARTHUR AND HIS KNIGHTS, Howard Pyle. The finest children's version of the life of King Arthur; brilliantly retold by Pyle, with 48 of his most imaginative illustrations. xviii + 313pp. 6⅛ x 9¼.
21445-1 Paperbound $2.50

THE WONDERFUL WIZARD OF OZ, L. Frank Baum. America's finest children's book in facsimile of first edition with all Denslow illustrations in full color. The edition a child should have. Introduction by Martin Gardner. 23 color plates, scores of drawings. iv + 267pp. 20691-2 Paperbound $2.50

THE MARVELOUS LAND OF OZ, L. Frank Baum. The second Oz book, every bit as imaginative as the Wizard. The hero is a boy named Tip, but the Scarecrow and the Tin Woodman are back, as is the Oz magic. 16 color plates, 120 drawings by John R. Neill. 287pp. 20692-0 Paperbound $2.50

THE MAGICAL MONARCH OF MO, L. Frank Baum. Remarkable adventures in a land even stranger than Oz. The best of Baum's books not in the Oz series. 15 color plates and dozens of drawings by Frank Verbeck. xviii + 237pp.
21892-9 Paperbound $2.25

THE BAD CHILD'S BOOK OF BEASTS, MORE BEASTS FOR WORSE CHILDREN, A MORAL ALPHABET, Hilaire Belloc. Three complete humor classics in one volume. Be kind to the frog, and do not call him names . . . and 28 other whimsical animals. Familiar favorites and some not so well known. Illustrated by Basil Blackwell. 156pp. (USO) 20749-8 Paperbound $1.50

EAST O' THE SUN AND WEST O' THE MOON, George W. Dasent. Considered the best of all translations of these Norwegian folk tales, this collection has been enjoyed by generations of children (and folklorists too). Includes True and Untrue, Why the Sea is Salt, East O' the Sun and West O' the Moon, Why the Bear is Stumpy-Tailed, Boots and the Troll, The Cock and the Hen, Rich Peter the Pedlar, and 52 more. The only edition with all 59 tales. 77 illustrations by Erik Werenskiold and Theodor Kittelsen. xv + 418pp. 22521-6 Paperbound $3.50

GOOPS AND HOW TO BE THEM, Gelett Burgess. Classic of tongue-in-cheek humor, masquerading as etiquette book. 87 verses, twice as many cartoons, show mischievous Goops as they demonstrate to children virtues of table manners, neatness, courtesy, etc. Favorite for generations. viii + 88pp. 6½ x 9¼.
22233-0 Paperbound $1.25

ALICE'S ADVENTURES UNDER GROUND, Lewis Carroll. The first version, quite different from the final *Alice in Wonderland,* printed out by Carroll himself with his own illustrations. Complete facsimile of the "million dollar" manuscript Carroll gave to Alice Liddell in 1864. Introduction by Martin Gardner. viii + 96pp. Title and dedication pages in color. 21482-6 Paperbound $1.25

THE BROWNIES, THEIR BOOK, Palmer Cox. Small as mice, cunning as foxes, exuberant and full of mischief, the Brownies go to the zoo, toy shop, seashore, circus, etc., in 24 verse adventures and 266 illustrations. Long a favorite, since their first appearance in St. Nicholas Magazine. xi + 144pp. 6⅝ x 9¼.
21265-3 Paperbound $1.75

SONGS OF CHILDHOOD, Walter De La Mare. Published (under the pseudonym Walter Ramal) when De La Mare was only 29, this charming collection has long been a favorite children's book. A facsimile of the first edition in paper, the 47 poems capture the simplicity of the nursery rhyme and the ballad, including such lyrics as I Met Eve, Tartary, The Silver Penny. vii + 106pp. 21972-0 Paperbound $1.25

THE COMPLETE NONSENSE OF EDWARD LEAR, Edward Lear. The finest 19th-century humorist-cartoonist in full: all nonsense limericks, zany alphabets, Owl and Pussycat, songs, nonsense botany, and more than 500 illustrations by Lear himself. Edited by Holbrook Jackson. xxix + 287pp. (USO) 20167-8 Paperbound $2.00

BILLY WHISKERS: THE AUTOBIOGRAPHY OF A GOAT, Frances Trego Montgomery. A favorite of children since the early 20th century, here are the escapades of that rambunctious, irresistible and mischievous goat—Billy Whiskers. Much in the spirit of *Peck's Bad Boy,* this is a book that children never tire of reading or hearing. All the original familiar illustrations by W. H. Fry are included: 6 color plates, 18 black and white drawings. 159pp. 22345-0 Paperbound $2.00

MOTHER GOOSE MELODIES. Faithful republication of the fabulously rare Munroe and Francis "copyright 1833" Boston edition—the most important Mother Goose collection, usually referred to as the "original." Familiar rhymes plus many rare ones, with wonderful old woodcut illustrations. Edited by E. F. Bleiler. 128pp. 4½ x 6⅜. 22577-1 Paperbound $1.25

TWO LITTLE SAVAGES; BEING THE ADVENTURES OF TWO BOYS WHO LIVED AS INDIANS AND WHAT THEY LEARNED, Ernest Thompson Seton. Great classic of nature and boyhood provides a vast range of woodlore in most palatable form, a genuinely entertaining story. Two farm boys build a teepee in woods and live in it for a month, working out Indian solutions to living problems, star lore, birds and animals, plants, etc. 293 illustrations. vii + 286pp.
20985-7 Paperbound $2.50

PETER PIPER'S PRACTICAL PRINCIPLES OF PLAIN & PERFECT PRONUNCIATION. Alliterative jingles and tongue-twisters of surprising charm, that made their first appearance in America about 1830. Republished in full with the spirited woodcut illustrations from this earliest American edition. 32pp. 4½ x 6⅜.
22560-7 Paperbound $1.00

SCIENCE EXPERIMENTS AND AMUSEMENTS FOR CHILDREN, Charles Vivian. 73 easy experiments, requiring only materials found at home or easily available, such as candles, coins, steel wool, etc.; illustrate basic phenomena like vacuum, simple chemical reaction, etc. All safe. Modern, well-planned. Formerly *Science Games for Children*. 102 photos, numerous drawings. 96pp. 6⅛ x 9¼.
21856-2 Paperbound $1.25

AN INTRODUCTION TO CHESS MOVES AND TACTICS SIMPLY EXPLAINED, Leonard Barden. Informal intermediate introduction, quite strong in explaining reasons for moves. Covers basic material, tactics, important openings, traps, positional play in middle game, end game. Attempts to isolate patterns and recurrent configurations. Formerly *Chess*. 58 figures. 102pp. (USO) 21210-6 Paperbound $1.25

LASKER'S MANUAL OF CHESS, Dr. Emanuel Lasker. Lasker was not only one of the five great World Champions, he was also one of the ablest expositors, theorists, and analysts. In many ways, his Manual, permeated with his philosophy of battle, filled with keen insights, is one of the greatest works ever written on chess. Filled with analyzed games by the great players. A single-volume library that will profit almost any chess player, beginner or master. 308 diagrams. xli x 349pp.
20640-8 Paperbound $2.75

THE MASTER BOOK OF MATHEMATICAL RECREATIONS, Fred Schuh. In opinion of many the finest work ever prepared on mathematical puzzles, stunts, recreations; exhaustively thorough explanations of mathematics involved, analysis of effects, citation of puzzles and games. Mathematics involved is elementary. Translated by F. Göbel. 194 figures. xxiv + 430pp. 22134-2 Paperbound $3.00

MATHEMATICS, MAGIC AND MYSTERY, Martin Gardner. Puzzle editor for Scientific American explains mathematics behind various mystifying tricks: card tricks, stage "mind reading," coin and match tricks, counting out games, geometric dissections, etc. Probability sets, theory of numbers clearly explained. Also provides more than 400 tricks, guaranteed to work, that you can do. 135 illustrations. xii + 176pp.
20338-2 Paperbound $1.50

CATALOGUE OF DOVER BOOKS

Mathematical Puzzles for Beginners and Enthusiasts, Geoffrey Mott-Smith. 189 puzzles from easy to difficult—involving arithmetic, logic, algebra, properties of digits, probability, etc.—for enjoyment and mental stimulus. Explanation of mathematical principles behind the puzzles. 135 illustrations. viii + 248pp.
20198-8 Paperbound $1.75

Paper Folding for Beginners, William D. Murray and Francis J. Rigney. Easiest book on the market, clearest instructions on making interesting, beautiful origami. Sail boats, cups, roosters, frogs that move legs, bonbon boxes, standing birds, etc. 40 projects; more than 275 diagrams and photographs. 94pp.
20713-7 Paperbound $1.00

Tricks and Games on the Pool Table, Fred Herrmann. 79 tricks and games—some solitaires, some for two or more players, some competitive games—to entertain you between formal games. Mystifying shots and throws, unusual caroms, tricks involving such props as cork, coins, a hat, etc. Formerly *Fun on the Pool Table*. 77 figures. 95pp.
21814-7 Paperbound $1.00

Hand Shadows to Be Thrown Upon the Wall: A Series of Novel and Amusing Figures Formed by the Hand, Henry Bursill. Delightful picturebook from great-grandfather's day shows how to make 18 different hand shadows: a bird that flies, duck that quacks, dog that wags his tail, camel, goose, deer, boy, turtle, etc. Only book of its sort. vi + 33pp. $6\frac{1}{2}$ x $9\frac{1}{4}$. 21779-5 Paperbound $1.00

Whittling and Woodcarving, E. J. Tangerman. 18th printing of best book on market. "If you can cut a potato you can carve" toys and puzzles, chains, chessmen, caricatures, masks, frames, woodcut blocks, surface patterns, much more. Information on tools, woods, techniques. Also goes into serious wood sculpture from Middle Ages to present, East and West. 464 photos, figures. x + 293pp.
20965-2 Paperbound $2.00

History of Philosophy, Julián Marías. Possibly the clearest, most easily followed, best planned, most useful one-volume history of philosophy on the market; neither skimpy nor overfull. Full details on system of every major philosopher and dozens of less important thinkers from pre-Socratics up to Existentialism and later. Strong on many European figures usually omitted. Has gone through dozens of editions in Europe. 1966 edition, translated by Stanley Appelbaum and Clarence Strowbridge. xviii + 505pp. 21739-6 Paperbound $3.00

Yoga: A Scientific Evaluation, Kovoor T. Behanan. Scientific but non-technical study of physiological results of yoga exercises; done under auspices of Yale U. Relations to Indian thought, to psychoanalysis, etc. 16 photos. xxiii + 270pp.
20505-3 Paperbound $2.50

Prices subject to change without notice.

Available at your book dealer or write for free catalogue to Dept. GI, Dover Publications, Inc., 180 Varick St., N. Y., N. Y. 10014. Dover publishes more than 150 books each year on science, elementary and advanced mathematics, biology, music, art, literary history, social sciences and other areas.